SpringerBriefs in Energy

SpringerBriefs in Energy presents concise summaries of cutting-edge research and practical applications in all aspects of Energy. Featuring compact volumes of 50 to 125 pages, the series covers a range of content from professional to academic. Typical topics might include:

- A snapshot of a hot or emerging topic
- A contextual literature review
- A timely report of state-of-the art analytical techniques
- An in-depth case study
- A presentation of core concepts that students must understand in order to make independent contributions.

Briefs allow authors to present their ideas and readers to absorb them with minimal time investment.

Briefs will be published as part of Springer's eBook collection, with millions of users worldwide. In addition, Briefs will be available for individual print and electronic purchase. Briefs are characterized by fast, global electronic dissemination, standard publishing contracts, easy-to-use manuscript preparation and formatting guidelines, and expedited production schedules. We aim for publication 8–12 weeks after acceptance.

Both solicited and unsolicited manuscripts are considered for publication in this series. Briefs can also arise from the scale up of a planned chapter. Instead of simply contributing to an edited volume, the author gets an authored book with the space necessary to provide more data, fundamentals and background on the subject, methodology, future outlook, etc.

SpringerBriefs in Energy contains a distinct subseries focusing on Energy Analysis and edited by Charles Hall, State University of New York. Books for this subseries will emphasize quantitative accounting of energy use and availability, including the potential and limitations of new technologies in terms of energy returned on energy invested.

More information about this series at http://www.springer.com/series/8903

Maxensius Tri Sambodo · Ahmad Helmy Fuady ·
Siwage Dharma Negara · Felix Wisnu Handoyo ·
Erla Mychelisda

Electricity Access and Community Welfare in Indonesia

Springer

Maxensius Tri Sambodo
Economic Research Center
Gedung Widya Graha LIPI
Indonesian Institute of Sciences (LIPI)
Kota Jakarta Selatan
Jakarta, Indonesia

Siwage Dharma Negara
ISEAS Yusof Ishak Institute
Singapore, Singapore

Erla Mychelisda
Economic Research Center
Gedung Widya Graha LIPI
Indonesian Institute of Sciences (LIPI)
Kota Jakarta Selatan
Jakarta, Indonesia

Ahmad Helmy Fuady
Research Center for Area Studies
Indonesian Institute of Sciences (LIPI)
Jakarta, Indonesia

Felix Wisnu Handoyo
Economic Research Center
Gedung Widya Graha LIPI
Indonesian Institute of Sciences (LIPI)
Kota Jakarta Selatan
Jakarta, Indonesia

ISSN 2191-5520 ISSN 2191-5539 (electronic)
SpringerBriefs in Energy
ISBN 978-981-16-3613-4 ISBN 978-981-16-3611-0 (eBook)
https://doi.org/10.1007/978-981-16-3611-0

This Springer imprint is published by the registered company Springer Nature Singapore Pte Ltd.
The registered company address is: 152 Beach Road, #21-01/04 Gateway East, Singapore 189721, Singapore

Foreword

Sustainable Development Goal (SDG) Number 7 mentions 'Ensure access to affordable, reliable, sustainable and modern energy for all by 2030.' This goal has a very important meaning for many countries facing the challenges of access to electricity, including Indonesia. Refer to the joint report of 'tracking SDG 7 the energy progress report 2020' by IEA, IRENA, United Nations, World Bank, and World Health Organization, said that Bangladesh, India, Indonesia, and the Philippines are well on track to reach full access before 2030.

However, this projection still needs to be supported by the better performance of all parties who are responsible for increasing access to electricity. Reflecting on Indonesia's success story in increasing access to electricity, it is very good to learn from the many valuable experiences discussed in this book. This book represents a synthesis of research findings on energy in remote rural areas in East Nusa Tenggara (NTT). The locus of the study is the finest example of an underdeveloped area in eastern Indonesia facing with serious energy poverty problems. For two consecutive years of research (2013 and 2014), we tried to measure carefully the impacts of access to electricity on the people's welfare. This book discusses some profound results of a field research to which decision makers need to pay their attention.

This book has been published in an Indonesian edition entitled 'Akses Listrik dan Kesejahteraan Masyarakat.' The authors attempt to update the data and discuss the study findings in the context of the current policies. Additional analysis and data updating were made as requested by book reviewers when published in the English version. In this way, this book can complement the publication in Indonesian.

We would like to take this opportunity to thank all those who have helped in preparing and carrying out the research as well as writing the final report. Our gratitude goes particularly to the subprogram coordinators of competitive research activities, resource persons, discussants, panelists, and participants in our focus group discussions or seminars for their valuable feedback and suggestions to make this study well developed. We hope the book becomes a significant reference to any policymakers and observers having a particular interest in the electricity sector in Indonesia.

Jakarta, Indonesia The Research Team

Publisher's Note

Lembaga Ilmu Pengetahuan Indonesia (LIPI) Press, as a scientific publisher, has the responsibility to produce high-quality scientific publications. The fulfillment of this statutory obligation is one of the publisher's roles in promoting the educational and intellectual life of the nation as mandated by Preamble of the 1945 Constitution.

This anthology of *Electricity Access and Community Welfare* presents research findings on the impacts of electricity access on the community welfare in some rural and remote areas in East Nusa Tenggara (NTT) where the research was conducted.

We hope that this book will aid the readers in finding sources of information on the impacts of electricity access on the welfare of communities living in remote rural areas. It is of our expectation that this book will also become a useful reference for researchers, policymakers, and observers working in the electricity sector. Finally, we thank all those who made this publication possible.

LIPI Press

Contents

About the Authors

Maxensius Tri Sambodo is Senior Researcher at the LIPI Economic Research Center and Lecturer at Universitas Nasional and Universitas Gadjah Mada. He earned a doctoral degree in public policy from the National Graduate for Policy Studies (GRIPS) in 2012. He is currently active in conducting researches in energy security. In 2014–2015, he was a visiting fellow at the Institute of Southeast Asian Studies (ISEAS) for completing his book entitled *From Darkness to Light: Energy Security Assessment in Indonesia's Power Sector*, published by ISEAS. He also co-authored other books, i.e., *the ASEAN Energy Market Integration (AEMI): From coordination to integration*, published by the ASEAN Studies Center, Chulalongkorn University, and *Kemiskinan Energi Listrik: Memberi Terang Wilayah Terpencil Indonesia*, published by LIPI Press.

Ahmad Helmy Fuady is Researcher at the Research Center for Area Studies, LIPI, Jakarta. He received a doctoral degree from the Faculty of Social and Behavioral Sciences, University of Amsterdam, Netherlands. His main research interests include development policy and political economy at the local, national, and global levels.

Siwage Dharma Negara is Researcher at the LIPI Economic Research Center, Jakarta. He obtained a doctoral degree in economics from the Department of Economics, University of Melbourne, Australia. His research interests include macroeconomic policies, energy security, industrial competitiveness, and human resources.

Felix Wisnu Handoyo is Researcher at the LIPI Economic Research Center, Jakarta. He got a bachelor's degree in economics from the Faculty of Economics and Business, Gadjah Mada University (FEB-UGM). His research focus areas are regional infrastructure development and energy security.

Erla Mychelisda is Researcher at the LIPI Economic Research Center, Jakarta. She has a bachelor's degree in economics from the Faculty of Economics, Jenderal Soedirman University, Purwokerto. Her research interests include energy security, management, and corporate finance.

Chapter 1
Introduction: Access to Electricity and Welfare

1.1 Background

Access to electricity is one of the basic preconditions for an improved quality of life and a more competitive economy in any society and any nation. No country in this world has ever reached high levels of development and well-being without adequate and sustainable access to electricity for its people. Economic growth will lead to higher energy demand and accelerate urbanization needs to be responded by modern energy supply. Thus, energy demand will continue to grow and almost two-thirds are driven by developing economies in the Asia—Pacific region (IEA, 2018). On the other hand, reliability of power supply has been challenged by power outages due to high risks on disaster (IEA, 2018).

Adequate access to electricity remains a major challenge in Indonesia. Although, according to the International Energy Agency (2018), the nation's electrification ratio (i.e. the total number of households with access to electricity compared to the total number of households), Indonesia was relatively higher than other ASEAN countries, in terms of electricity consumption, Indonesia was relatively lower than other countries. This implies that low electricity consumption reflects a access constraint and lack of capacity for productive used (Table 1.1).

In addition to lack of access, inter-provincial disparities in provision of electricity are still a major issue in Indonesia. In 2018, the electrification ratio in areas outside Java is relatively low, which was 90.54% as compared to nearly 100% in Java (PLN, 2018). The situation is more unfortunate for eastern Indonesia, especially Province of Papua (42.5%) and East Nusa Tenggara (56.34%) (see Fig. 1.1).

As indicate by the linier trend, there is evidence that low access to electricity correlates with lower economic growths and higher poverty rates. However, it does not mean causation. For example, electricity power consumption per capita in East Nusa Tenggara Province (NTT) is only 182.55 kWh per capita (PLN, 2019), which is well below the national average of 740 kWh per capita, or only 25% of the national average electricity consumption. By September 2019, the proportion of total population living below poverty line in NTT reached 20.6% or the third highest after Papua

M. T. Sambodo et al., *Electricity Access and Community Welfare in Indonesia*, SpringerBriefs in Energy, https://doi.org/10.1007/978-981-16-3611-0_1

Table 1.1 Comparison of electricity access in developing Asia

Country	Electrification ratio (%)	Electricity consumption (kWh)
Brunei	>95	10,291
Cambodia	72	271
Laos	95	–
Indonesia	98	812
Malaysia	>5	4652
Myanmar	43	215
Philippines	>95	696
Singapore	>95	8845
Thailand	>95	2539
Viet Nam	>95	1424

Source IEA, World Energy Outlook (2018)

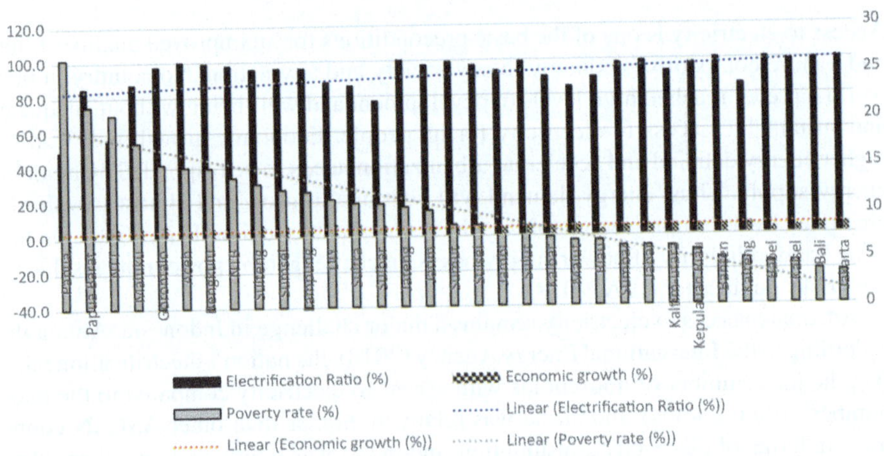

Fig. 1.1 Electrification ratio, poverty, and economic growth by province in 2019. *Source* BPS and PLN (2019)

and West Papua (BPS, 2019a). The economic growth of NTT in 2018 was about 5.2%, which was still low as compared to the Java economic growth of 5.67% (BPS, 2019b). Consequently, the slow pace of economic growth impeded the job creation process and the poverty reduction program in the province.

The Indonesian government is fully aware of the importance of expanding electricity access to rural areas. Law No. 30/2007 on Energy specifically states that the central and regional governments are obliged to provide financial assistance for the availability of electricity to poor people, to build energy infrastructures in underdeveloped regions, to improve electricity access in remote and border areas, and to develop rural electrification. The Law also confirms that the central and regional governments are responsible to guarantee the availability of energy for people's needs.

> …….improving accessibility to energy for the people who are less wealthy and/or who live in remote areas to bring about just and equal welfare and prosperity for the people by: 1) providing assistance to increase the availability of energy for less wealthy people; 2. building energy infrastructures in underdeveloped regions in order to reduce disparities among regions. (Article 3 [f])

Moreover, Law No. 30/2009 on Electricity mentions that:

> Power supplies shall be under the control of the state, which supplies shall be provided by the Government and the regional governments under the principle of regional autonomy (Article 3 [1]).

That is, the local governments under regional autonomy bear responsibility (with the support of the central government) to actively improve access to electricity in their areas. The regional governments can play a role by establishing the Region-Owned Enterprise (*Badan Usaha Milik Daerah,* BUMD) targeting specifically the electricity sector. They can also give opportunities to private business entities and cooperatives to pursue the electricity sector development in their regions (see Article 11 [3], Law No. 30/2009). Where no BUMDs, private entities, or cooperatives are able to provide electricity in certain areas, the government must give an order to the State-Owned Enterprise (*Badan Usaha Milik Negara,* BUMN), which is in this case the State Electricity Enterprise (*Perusahaan Listrik Negara,* PLN), to carry out these functions (see Article 11 [4], Law No. 30/2009).

The Laws No. 30/2007 and No. 30/2009 gave rise to a number of electricity programs in various parts of Indonesia. Since 2006, the government, through PLN, has been introducing the accelerated program phase I of 10, 000 MW continued with the phase II of 10.000 MW to be completed by the end of 2014. While the first phase gave priority to environmentally harmful coal-fired power plants, the second prioritized eco-friendly geothermal ones since 2010. The two-phase accelerated program showed lower-than-expected levels of realization because of many obstacles encountered.

In addition to the accelerated power development program, the government has been seeking to improve access to electricity through the so-called "One Day One Million Connection" program. Introduced by PLN in 2010, the movement has been successfully reducing the queues of potential electricity customers. For example, PLN was able to fulfill 62.7% of electrical connection requests in 2010 and it increased to about 99% in 2018 (Fig. 1.2). However, in 2012 and 2013, the high rate of queues resumed as demand for electrical connections was beyond the PLN's production capacity.

The government has been also striving to boost the utilization of renewable and renewable energy through a 100% solar energy power program in 100 islands (Yuli, 2011) and construction of hydroelectric power plant, such as mini-, and micro- hydro and geothermal power plants (*Pembangkit Listrik Tenaga* PLTP). In spite of all these efforts, the government has to face big challenges and constraints in securing people's access to electricity, especially in the rural, border, and outer areas.

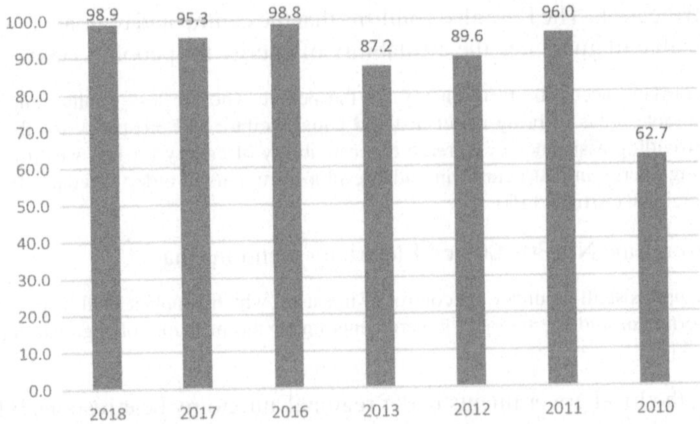

Fig. 1.2 PLN Electricity demand fulfillment, 2010–2018 (in percentage). *Note* Basically, every year there are new demand for electricity connection, but some can be connected, and the rest need to wait or cancel. The electricity demand fulfillment = (new connection / (new demand − cancel)) × 100 . *Source* PLN, various year

This book attempts to address the challenges and obstacles in the electricity sector development in rural areas. With a focus study on remote rural areas of NTT, it explores some of the major problems in having a fruitful electricity development for rural communities. The hope is that the people will be able to meet their energy needs, to improve their capabilities of preserving sustainable power installations, to develop a more competitive renewable energy source, and to tackle their geographical barriers in the face of minimal infrastructures.

The book also scrutinizes the impacts of electricity on the communities amid the growth of investment activities, productive community businesses, and household income and savings. It particularly seeks to connect access to electricity with household expenditure patterns and behaviors. The chapters will address some important questions: does access to electricity alter household spending patterns and behaviors, such as on kerosene and firewood as well as on education and health? Does it stimulate household productive activities and increase access to information? Finally, the book examines the strategies of electricity sector development in support of poverty alleviation programs in rural areas.

This book is divided into five chapters. The first chapter described the goals, frameworks, methodologies, and findings of the research. The second explained the potential energy resources at the research sites, including Tantong Hamlet, Lungar Hamlet and Damu Hamlet, Satarmese Sub-District, Manggarai District, and NTT Province. The third provided an analytical account on the institutional models for the development of PLTP. The fourth examined the impacts of electricity access on the community welfare based on the panel data from 2013 to 2014. The fifth focused on the development agenda of the rural electricity sector, studying the possible synergy between the rural development programs and the electricity sector improvement. The end on of the chapter provided conclusion and policy recommendations.

1.2 Framework

It is unquestionable that economic growth plays a significant role in reducing poverty problems in any region. The high economic growth would require the availability of excellent infrastructures, one of which is access to electricity. Adequate access to electricity is a basic precondition for the acceleration of economic growth. As seen from Fig. 1.3, development of access to electricity becomes the starting point for improving quality of live.

Previous studies showed that there is a link between access to electricity and people's welfare (see Al Mohtad, 2006; Kanagawa & Nakata, 2008; Munasinghe, 1988; Peng & Pan, 2006; Reiche et al., 2000). Reiche et al. (2000), for example, found that rural electrification programs can effectively improve living standards (such as public amenities and services); they can reduce the negative effects of conventional energy use (firewood and coal) on health and environment; and they can also directly

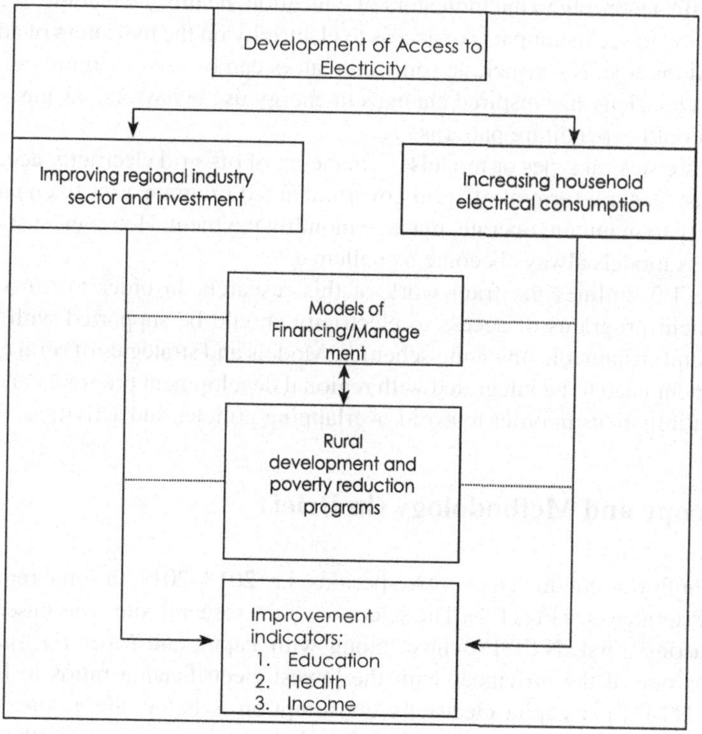

Fig. 1.3 Framework of the research. *Source* Sambodo et al., (2013)

or indirectly create more employment opportunities and enhance productivity. Similarly, a research by Kanagawa and Nakata (2008) in one of the poorest states in India also demonstrated that access to electricity has a direct and indirect connection with the indicators of health, education, income, and environment.

Kanagawa and Nakata (2008) asserted further that the impacts of access to electricity are under the influence of the conditions of infrastructures, power capacity, government policy, and international cooperation. Access to electricity, for instance, has a positive impact on socio-economic indicators, such as a rise in the levels of education, health, and other variables. In line with Kanagawa and Nakata (2008), Khandker et al. (2013) found that the benefits of electrification programs to people's welfare can be measured by using three outcome indicators, i.e. education, income, and public health. Their research has also proven that it is hard to assess the direction and magnitude of causality between electrification programs and outcome indicators due to a complex linkage of different electrical devices, outputs and outcomes. They pointed out that access to electricity allows households to use various electronic devices, such as electric lights, radios, televisions, refrigerators, cooking appliances and so on. As a result, all the electrical tools generate a variety of output, such as lighting, information access, convenience, better food storage, productive activities, and more efficient cooking methods.

The development of access to electricity is aimed at improving the quality of people's life according to the indicators of education, health and income. It may take a long period to see the impacts of access to electricity on the indicators of education, health and income. Nevertheless, some indicators can be used to figure out whether access to electricity has inspired changes in energy use behaviors, business models, and household expenditure patterns.

There are several types or models of financing of off-grid electricity access in the rural areas. General form of off-grid government led program aims to enhance local community to maintain, operate, manage monthly payment. However, sustainability of business models always become a challenge.

Figure 1.3 outlines the framework of this research. In order to run smoothly, development programs of access to electricity should be supported with efficient, reliable, and sustainable financing schemes. Models and strategies of rural electricity development need to be integrated with regional development programs and poverty reduction initiatives in order to avoid overlapping policies and activities.

1.3 Scope and Methodology (in Brief)

The research was conducted over two periods, i.e. 2013–2014, in some remote areas of NTT Province (see Fig. 1.4). The selection of the research sites was based on three main reasons. First, NTT Province, along with Papua and *Nusa Tenggara Barat* (NTB), is one of the provinces with the lowest electrification ratios in Indonesia. Second, NTT's per capita electricity consumption is below the national average, undoubtedly having a positive correlation with the high poverty rate in the province. Third, NTT Province has a great potential of renewable energy resources, such as

Fig. 1.4 Research sites in Hamlets of Tantong, Damu and Lungar, Sub-district of Satarmese, District of Manggarai, NTT. *Source* US Dept. of State Geographer, Accessed from Google Earth (2016 and 2020)

solar, biofuel, wind, and geothermal. It is interesting that even though the province has a Geothermal Power Plant (PLTP) in Ulumbu, many surrounding villages are still denied access to electricity.

The research team selected two groups of people to assess the impacts of access to electricity on the people's economic behaviors and patterns. The first group is households with access to electricity (the treatment group), while the second is households without access to electricity (the comparison group). The indicators used to measure behavioral changes in the household energy consumption are expenditure on kerosene (a source of fuel for lighting), expenditure on firewood (wood fuel for cooking), and expenditure on foods and non-foods. The indicators used to appraise changes in household welfare are income levels, working hours or days, study hours among school-age children, and total household expenditure.

In the first year of the research (2013), the team carried out baseline surveys to find out the socio-economic conditions of all households in Tantong Hamlet, Damu Hamlet, Lungar Hamlet, and Mesir Hamlet. During the survey period, the population of Wewo Village, consisting of Tantong Hamlet and Damu Hamlet, was 2132 people and 486 households, and Lungar Village was 1427 people and 285 households. Wewo Village was chosen because of the information that one of its hamlets, i.e. Tantong Hamlet, will have access to electricity in 2013, which came to actually happen in April 2014.

The other two hamlets, Lungar Hamlet and Damu Hamlet were studied for two reasons. The first is that Lungar Hamlet is an area surrounded by a range of hills separating it from Tantong Hamlet and Damu Hamlet. Because of this topography, it is difficult for the PLN's cable transmitted electricity services to reach the area. Under this condition, therefore, the households without access to electricity in Lungar Hamlet can be taken as a comparison group. On the other hand, Damu Hamlet and Tantong Hamlet are geographically close to each other and most of their residents are farmers. Damu Hamlet has a somewhat better infrastructure thank to the availability of road and electricity access. This means also that Damu Hamlet did not actually fit into the category of either the treatment group or the comparison group. The research team decided to keep gathering data from Damu Hamlet due to its close proximity to Tantong Hamlet. Thus, the spillover effects are well-studied.

Table 1.2 shows us the results of baseline surveys at the three research sites. In 2013, most households in Tantong Hamlet did not have access to the PLN's electricity. Twenty-four households who have access to electricity are still using lamps for

Table 1.2 Conditions of Household's Access to Electricity Based on Surveys in 2013 and 2014

Status	Tantong		Damu		Lungar	
	2013	2014	2013	2014	2013	2014
With access to electricity	24	57	44	45	120	134
Without access to electricity	36	5	0	0	83	102
Total number of households	60	62	44	45	203	236

Source Authors (primary data)

lighting. Some households got the supply of electric currents from wires stretched to the households that already have access to the PLN's electricity in Damu Hamlet. Due to the insufficient power supply, some households still have to use kerosene for lighting. At the same time, electricity is already accessible to some households in Lungar Hamlet. For lighting, people in the hamlet depend on electricity sources provided by the Super Extra Energy-Saving (SEHEN) program that utilizes Sun's rays or solar panels.

The survey was carried out over more than 300 heads of families (KK) in all three hamlets during 2013–2014 and it covered the village population. There were 307 families registered in 2013 and the number increased to 344 in 2014. The data collection process involved about 20 local field assistants. The baseline data was obtained throughout May–June 2013, while the midline one throughout June-July 2014.

1.4 Research Findings

The analysis on the baseline data showed that access to electricity correlates with a decrease in the household expenditure on kerosene as a fuel for lighting. In general, the households spend for foods the money they "saved" from their reduced kerosene expenditure. The baseline and midline data have not yet indicated the impact of access to electricity on productivity in the form of longer working hours. However, the students spend longer hours for study after getting access to electricity.

Based on the condition in the research field, it seemed that the sustainability of rural electrification programs remain questionable. Today, the various programs of rural electrification tend to focus on improving the physical facilities while neglecting the managerial and sustainable aspects. The SEHEN program, which was expected to become more sustainable than any other electricity programs run by the ministries or other organizations, had to encounter multiple obstacles after being implemented. One of the indicators is that unpaid electric bills among the SEHEN customers were at a high rate, causing PLN to lose money. The government programs were also lack of coordination and synergy, as strikingly evident from a number of overlapping programs undertaken by the ministries. The Law on Electricity did not stand alone in regulating the supply of electricity. For example, the implementation of geothermal development should take into account the regulations pertaining to the forestry sector. However, the new Geothermal Law (Law No. 21/2014) now allows indirect geothermal utilization in the protected, production, and conservation forest areas.

Law No. 30/2009 on Electricity has not yet answered the aspirations of local governments to become more involved in providing access to electricity. The Geothermal Law is considered incompatible with the spirit of regional autonomy, indirectly stirring a conflict between the geothermal entrepreneurs. At a practical level, it seems that the regional governments have been unready to take part in the management of electricity. The District of Manggarai was an example, where the regional government has not prepared any regulations on electricity as well as its managing institution (i.e. BUMD). In addition, the promise of high potential revenues from the utilization of geothermal energy would make the regional governments reluctant to easily let the PLN take over the management of electricity.

References

Al Mohtad, I. (2006). Remote area power supply systems (RAPSS). *Himala- yan Small Hydropower Summit*. October 12–13, 2006.
Central Agency of Statistics (BPS). (2019a). *Pertumbuhan Ekonomi Indonesia Triwulan IV-2018*.
Central Agency of Statistics (BPS). (2019b). *Profil kemiskinan di Indonesia September 2018*.
International Energy Agency (IEA). (2018). *World energy outlook 2018*. International Press.
Kanagawa, M., & Nakata, T. (2008). Assessment of access to electricity and the socio-economic impacts in rural areas of developing countries. *Energy Policy, 36*, 2016–2029.
Khandker, S. R., Barnes, D. F., & Samad, H. A. (2013). Welfare impacts of rural electrification: A panel data analysis from Vietnam. *Economic Development and Cultural Change, 61*(3), 659–692.
Munasinghe, M. (1988). Rural electrification: International experience and policy in Indonesia. *Bulletin of Indonesia Economic Studies, 24*(2), 87–105.
Peng, W., & Pan, J. (2006). Rural electrification in China: History and institution. *China & World Economy, 14*(1), 71–84.
PLN. (2014). Status, target, dan tantangan program elektrifikasi nasional dan investasi energi terbarukan. A paper presented at FGD conducted by LIPI on July 21, 2014.
PLN. (2018). *Statistik PLN 2018*. Jakarta: PT PLN.
Reiche, K., Covarrubias, A., & Martinot, E. (2000). *Expanding electricity access to remote areas: Off-grid rural electrification in developing countries*. World Power.
Republic of Indonesia. (2007). Energi. *Undang-undang No. 30 Tahun 2007 tentang Energi*.
Republic of Indonesia. (2009). Ketenagalistrikan. *Undang-undang No. 30 Tahun 2009 tentang Ketenagalistrikan*.
Republic of Indonesia. (2014). Panas bumi. *Undang-undang Republik Indonesia No. 21 Tahun 2014 tentang Panas Bumi*.
Sambodo, M. T., Fuady, A. H., Negara, S. D., Handoyo, F. W., & Mychelisda, E. (2013). *Laporan Penelitian Model dan Strategi Pengembangan Sektor Ketenagalistrikan di Daerah dalam Upaya Pengentasan Kemiskinan*. Jakarta: Pusat Penelitian Kependudukan.
Yuli. (2011, 9 April). Bangun listrik tenaga surya di 100 pulau. *Kompas. com*. Accessed on May 2013 from http://bisniskeuangan.kompas.com/read/2011/04/09/07431522/bangun.listrik.tenaga.surya.di.100.pulau

Chapter 2
Potential of Energy Resources in East Nusa Tenggara

2.1 The Energy Sector in NTT Province: An Overview

NTT Province comprises 20 districts and one city located in the seven major islands, i.e. Sumba, Timor, Flores, Alor, Lembata, Rote, and Sabu (BPS, 2019). The archipelagic landscape of NTT has brought some challenges to the province in providing electricity to its communities; thus, off-grid electrification with renewable energy sources was widely developed particularly to supply power to the small islands.

The growth rates of the economy are one of the factors that increase the energy demands in NTT. The electricity used in the province has rose to 8.4% in 2018, even higher than the national need, which only 5% (Lewokeda, 2019). The increasing number is dominated by industry such as cement that grows around 14%, and businesses such as hotels, entertainment, and banking that grows 10%. There was also the growth of electricity usage from the multi-use segment in the form of special services for project development or investment needs, for example, the construction of the Trans Mart shopping center in Kupang City that reached 2.7 MW (Lewokeda, 2019). Almost all the entire NTT region experienced growth in electricity usage, but few areas grew higher, such as in Labuan Bajo that reached more than 20% due to tourism growth. Meanwhile, in Belu regency, the use of the industry has grown quite significantly, along with the presence of various government development programs in the country's border areas (Lewokeda, 2019).Another factor that raises the energy demand in NTT is the growth of the population. Over the period 2011 to 2017, the NTT's population grew around 1.6–1.75% per annum and will likely continue to experience a positive figure (BPS Provinsi NTT, 2018b). The total population was around 4.6 million in 2009 and increased to around 5.2 million in 2017 (BPS Provinsi NTT, 2018a). Currently, the per capita energy consumption in NTT Province is less than the national average. Therefore, there is an urgent need for the development of the energy sector in the province.

M. T. Sambodo et al., *Electricity Access and Community Welfare in Indonesia*, SpringerBriefs in Energy, https://doi.org/10.1007/978-981-16-3611-0_2

Table 2.1 Load factor, capacity factor, and demand factor, 2005–2013

Year	Load factor	Capacity factor	Demand factor
2005	75.48	52.51	47.48
2006	64.15	48.00	55.53
2007	59.60	64.47	47.10
2008	80.77	52.62	28.12
2009	76.37	53.71	29.81
2010	77.78	55.90	29.56
2011	78.53	55.67	28.37
2012	79.18	51.96	27.54
2013	80.04	54.72	26.50

Note Capacity factor is the ratio of the gross energy production in one year to the installed power capacity multiplied by 8760 hours (PLN, 2013a)
Demand factor is the ratio of the maximum demand to the real energy power (PLN, 2013a)
Source PLN (2013a)

Moreover, the electrical load or load factor will most likely keep increasing as the economy and population are growing in NTT. The electrical load factor is a comparison between the average load in a specified time period and the peak load taking place in that period (PLN, 2013a). At present, the electrical load factor in NTT has not yet been optimized (see Table 2.1). The energy needs in the future should have been ideally answered by having more eco-friendly and affordable power plants. However, not every power plant has such a characteristic. Thus, the energy sources used in NTT has been mixed between fossil and renewables, where hydropower and geothermal, are usually opted for fulfilling the baseload demands, while diesel-fueled generators are for the peak load demands (Table 2.2).

In the 2013–2022 Electricity Supply Business Plan, PLN plans to develop power plants fueled by fossil and new renewable energy sources in order to answer the energy needs in NTT (Table 2.3). Geothermal Power Plant (*Pembangkit Listrik Tenaga Panas Bumi,* PLTP) and Hydro Power Plant (*Pembangkit Listrik Tenaga Air,* PLTA) are two renewable energy sources to be potentially developed. Unfortunately, the development progress until now has remained slowed. The power plant capacity owned by PLN which is being developed on the island of Flores, as an existing geothermal icon island, has only reached 13 MW, still far from the 115 MW target (ANTARA News, 2019).

Moreover, to increases the electrification ratio, besides expanding the power plant development and capacity, PLN investment plan also declares an intention to expand the electricity transmission and distribution channels. The transmission channels are used to deliver electrical energy from the power plant to various regions. By raising the height of voltage (up to 70 kV, for instance) channeled through the transmission lines that interconnect the cities, electricity losses can be avoided. Because these cities are in general situated at the center of the islands, the coastal areas might suffer

Table 2.2 Projected energy needs in NTT

Year	Economic growth (%)	Selling (GWh)	Production (GWh)	Peak load (MW)	Consumer
2013	5.6	640	716	140	838,382
2014	5.9	734	830	160	922,977
2015	6.0	802	946	175	1,013,817
2016	6.0	876	1064	190	1,097,947
2017	6.0	958	1170	207	1,195,962
2018	6.0	1050	1288	226	1,234,115
2019	6.0	1153	1424	247	1,254,184
2020	6.0	1267	1568	271	1,296,117
2021	6.0	1394	1739	296	1,337,533
2022	6.0	1536	1921	325	1,374,933
Growth	5.9	10.5%	11.6%	9.6%	7.3%

Source PLN (2013a)

a lack of electricity unless the transmission channels are appropriately taken to the given course.

The expansion of transmission channels coincided with the extension of substations (Table 2.4). The substation has a step-down transformer through which a high voltage is lowered to a medium rate of 20 kV. The medium voltage is then carried over into a Medium Voltage Network (JTM). The establishment of power plants, transmission channels and substations has also been accompanied by the construction of distribution channels that will transport electricity to the consumers. By doing so, the security and efficiency of electrical energy can be attained.

2.2 Energy Potentials in NTT

Like many other small regions in Indonesia, NTT is still incapable of fully and independently meeting its energy needs. The fact showed that almost 96% of the power plants in NTT were diesel-fueled in 2013 (PLN, 2013b) and the diesel oil has to be imported from other provinces. As Likadja (2012) argue that NTT is lack of fossil energy. However, NTT has enormous potential of alternative energy, such as geothermal, hydro, solar, wind and ocean waves. If the utilization of renewable energy in NTT has run well, NTT will likely solve its deficit problem.

Figure 2.1 shows that the use of geothermal energy has exerted a positive impact on reducing diesel consumption. In 2013, the total saving rate was IDR 75.5 billion, and it tends to go higher provided that the construction of geothermal power plants

Table 2.3 Power plant development plan in NTT

No.	Project	Assumed developer	Type	Capacity (MW)	COD
1	Ende (FTP 1)	PLN	PLTU	2 × 7	2013
2	Kupang (FTP 1)	PLN	PLTU	2 × 16.5	2013
3	Ndungga	PLN	PLTM	1.9	2014
4	Ulumbu (ADB)	PLN	PLTP	5	2014
5	PLTM Tersebar NTT	PLN	PLTM	3.35	2014/15
6	Rote Ndao	PLN	PLTU	2 × 3	2015
7	Alor	PLN	PLTU	6	2015
8	Maumere	PLN	PLTU	2 × 10	2016
9	Atambua	PLN	PLTU	4 × 6	2016
10	Wae Rancang I	PLN	PLTA	10	2017
11	Wae Rancang II	PLN	PLTA	6,5	2018
12	Kupang	Private	PLTU	30	2015/16
13	PLTM Tersebar NTT	Private	PLTM	9.15	2014/15/16
14	Mataloko (FTP 2)	Private	PLTP	5	2018
15	Atadei (FTP 2)	Private	PLTP	5	2017
16	Sokoria (FTP 2)	Private	PLTP	3 × 5	2018/19/20
17	Oka Ile Ange (FTP 2)	Private	PLTP	10	2020
18	Kupang Peaker	*Unallocated*	PLTMG	40	2016
19	Maumere Peaker	*Unallocated*	PLTMG	20	2017
20	Timor 1	*Unallocated*	PLTU	2 × 15	2018/19
21	Kupang Peaker 2	*Unallocated*	PLTMG	20	2018
22	Bukapiting	*Unallocated*	PLTP	5	2020
23	Kupang Peaker 2	*Unallocated*	PLTMG	10	2022
24	Timor 2	*Unallocated*	PLTU	25	2022
25	Larantuka	*Unallocated*	PTMPD	2.4	2015
26	Sumba	*Unallocated*	PLT-Biomass	1	2016
27	Adonara	*Unallocated*	PTMPD	2 × 1.2	2016
28	Mataloko 3	*Unallocated*	PLTP	5	2019
29	Ulumbu 3	*Unallocated*	PLTP	5	2019
30	Mataloko 4	*Unallocated*	PLTP	5	2020
31	Ulumbu 4	*Unallocated*	PLTP	5	2021
32	Sokoria 4	*Unallocated*	PLTP	5	2022
	Jumlah			385	2013–2022

COD commercial operating date
Unallocated means the need for additional capacity that has not been committed but referred to as PLN or IPP's unallocated project
Source PLN (2013a)

Table 2.4 Development plan of substation 70 kV in NTT 2014–2022

No.	Name of substation	Voltage	New/Extension	Capacity (MVA)	COD
1	Ropa	70/20 kV	New	5	2014
2	Bolok	70/20 kV	New	20	2014
3	Maulafa	70/20 kV	New	30	2014
4	Ende	70/20 kV	New	10	2014
5	Maumere	70/20 kV	New	10	2014
6	Naibonat	70/20 kV	New	20	2014
7	Nonohonis	70/20 kV	New	20	2014
8	Kefamenanu	70/20 kV	New	20	2014
9	Atambua	70/20 kV	New	20	2014
10	Atapupu	70/20 kV	New	10	2014
11	Labuan Bajo	70/20 kV	New	20	2014
12	Bajawa	70/20 kV	New	20	2015
13	Ruteng	70/20 kV	New	20	2015
14	Maumere	70/20 kV	Extension	20	2015
15	Ende	70/20 kV	Extension	20	2015
16	Maulafa	70/20 kV	Extension	30	2015
17	Naibonat	70/20 kV	Extension	20	2016
18	Bolok	70/20 kV	Extension	20	2016
19	Maulafa	70/20 kV	Extension	30	2017
20	Atambua	70/20 kV	Extension	20	2019
21	Larantuka	70/20 kV	New	20	2019
22	Ropa	70/20 kV	Extension	20	2019
23	Ende	70/20 kV	Extension	20	2019
24	Bajawa	70/20 kV	Extension	20	2020
25	Bolok	70/20 kV	Extension	20	2020
26	Maumere	70/20 kV	Extension	20	2020
27	Maulafa	70/20 kV	Extension	30	2021
28	Nonohonis	70/20 kV	Extension	20	2022
		Total		555	2014–2022

SUTT high voltage air channels
Source PLN (2013a)

in NTT begins to move more quickly. The acceleration will result in a reduction not only in electricity subsidies but also in CO_2 emissions from the power generation sector.

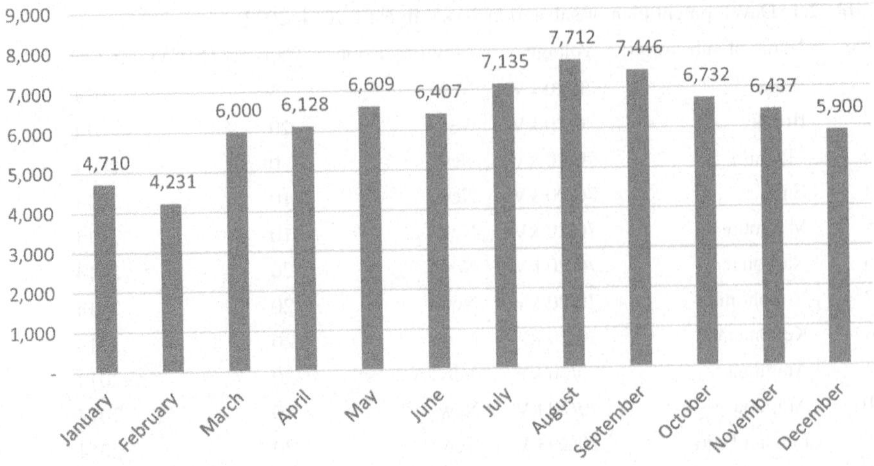

Fig. 2.1 Fuel saving figures by the operation of the PLTP Ulumbu 2013 (in Million Rupiah). *Source* PLN Ruteng Substation (2014)

2.3 Knowing the Potential and Power Generation Sources in NTT Province

2.3.1 Fossil Power Plant

The local State-Owned Electricity Enterprise (*Perusahaan Listrik Negara,* PLN) has made various investments to handle the needs for electricity in NTT. The Diesel Fueled Power Plant (*Pembangkit Listrik Tenaga Diesel* PLTD), for example, has been undertaking a major task of relocating 2 × 2.5 MW diesel engines in Lembata of East Flores Branch (Januwarsono, 2010). This relocation usually aims at optimizing electrical power. Meanwhile, the PLTDs that have been currently operating in Manggarai District include PLTD Waso 4154 kW), PLTD SR Borong (1300 kW), PLTD SR Reo (1325 kW), PLTD SR Pota (400 kW), and PLTD Benteng Jawa (200 kW) (PT PLN Ruteng Substation, 2014).

Coal is one of the energy sources in NTT for which the Provincial Government has issued Regulation No. 8/2010 on Management of Coal and Mineral Mining (Government Province of NTT, 2010). The rule encompasses various aspects related to the scopes of management, authority, mining material classification, mining area planning, mining area proposal, mining area change, and other technical features.

Currently, PLN NTT is still dependent on Steam Power Plant (*Pembangkit Listrik Tenaga Uap,* PLTU) 1 in Ende for 2 x 7 MW and PLTU 2 in Kupang for 2 × 16.5 MW. Januwarsono (2010) said that several new steam power plants would be built to elevate the power supply in NTT. They include:

1. Kupang Branch:
   1. Coal Power Plant of 4 × 6 MW in Atambua
   2. IPP Coal Power Plant of 2 × 15 MW in Kupang
   3. Coal Power Plant of 2 × 3 MW in Rote
   4. Coal Power Plant of 2 × 3 MW in Alor.
2. West Flores Branch:
   1. Coal Power Plant of 2 × 3 MW in Labuan Bajo
   2. Coal Power Plant of 2 × 7 MW Coal in Ropa, Ende District.
3. East Flores Branch:
   1. Coal Power Plant of 2 × 4 MW in Larantuka.
4. Sumba Branch:
   1. IPP Coal Power of C2 × 4 MW in East Sumba District.

2.3.2 Hydro Power Plant

NTT Province is a hilly area with 40 rivers the length of which varies from 25 to 118 km. These rivers are potential places for the construction of hydro power plants. The hydro energy potential in NTT is amounting to 20,705 kW (Table 2.5).

Unfortunately, this potential has not been fully exploited for a Hydro Power Plant (PLTA). The water in NTT has the potential utility for Mini-Hydro Power Plant (PLTM) while the residual energy can be further developed, through a cascade system, for a Micro-Hydro Power Plant (PLTMh) or a Pico-Hydro Power Plant

Table 2.5 Hydro Energy Potential in NTT

No.	District	Number of potential location	Estimated capacity (kW)
1	West Sumba	5	1808
2	East Sumba	4	222
3	Sikka	3	748
4	Manggarai	7	7802
5	TTU	1	48
6	TTS	7	7489
7	Ngada	8	2021
8	Ende	3	260
9	Belu	2	272
10	Alor	2	34
	Total	42	20,705

Source Likadja (2014)

(PLTPh) which is the simplest among other the hydro power plants. Moreover, NTT is home to five hydro power plants with a capacity of 1.14 MW (PLN, 2013b). Januwarsono (2010) added that the construction of hydro power plants in NTT is distributed to three branches:

1. West Flores Branch:
    1. IPP PLTM of 1 × 400 kW IPP PLTM in Ngada District
    2. IPP PLTM of 2 × 500 kW IPP PLTM in East Manggarai District.
2. Eastern Flores Branch:
    1. IPP PLTM of 2 × 400 kW in Lembata District.
3. Sumba Branch:
    1. IPP PLTM of 2 × 800 kW in Wanokaka, West Sumba District
    2. IPP PLTM of 3 × 350 kW in Kambaniru, East Sumba District.

2.3.3 *Ocean Current Power Plant*

NTT Province is traversed by the Indonesia Throughflow (*Arus Lintas Indonesia*, Arlindo), which refers to a route of currents that carries water masses from the Pacific Ocean toward the Indian Ocean (Fig. 2.2). Arlindo occurs due to the southeast trade wind in the south Pacific regions of Indonesia. The trade wind made the surface waters in the tropical part of the western Pacific Ocean higher than in the eastern

Fig. 2.2 Indonesia through flow. *Note* Arlindo is utilized to detect: El Nino, La Nina, and Fish Migration. *Source* Setiawan et al. (2011)

Indian Ocean, causing a flow from the Pacific Ocean to the Indian Ocean. During the southeast monsoon, Arlindo is in general stronger than the northwest monsoon. Arlindo renders Indonesia a potential of 5.6–9 terawatt (Setiawan et al., 2011).

The research conducted by the Agency for the Assessment and Application of Technology (Badan Pengkajian dan Penerapan Teknologi, BPPT) found that the straits in NTT and NTB have the electricity potential of 300 MW. There are at least ten straits with heavy currents that are prospective to produce electricity, such as Alor, Alas, Boleng, Flores, Linta, Lamakera, Molo, Pantar, and Sape (Electricity of Indonesia, 2013).

However, the ocean current power in NTT has been in the stage of prototype development and has not yet reached its optimum utilization (BPPT, 2010). The construction of offshore facilities required by a current power plant still becomes a considerable barrier. The building is needed to channel further the energy through submarine cables, the installation of which should be placed carefully in locations that are safe or not disruptive to the shipping lane. In addition, the establishment of a current power plant also necessitates water-resistant sea materials which cost much money.

In 2018, the government planned to set up a feasibility study to build an ocean current power plant in the Strait of Larantuka, East Flores due to its strong current potential. The project was carried out by a consortium from the Netherlands, Tidal Bridge BV and PJB in a Joint Venture with Tidal Bridge Indonesia. Tidal Bridge assumes that the average sea current speed of the Larantuka Strait is 3.5 m/s, the installed capacity of each turbine is 16 MW with the energy generated effectively of 6 MW. With the assumption of installing 5 turbines, the energy will generate an average of 30 MW (Kementerian ESDM RI, 2018).

2.3.4 Wind Power Plant

Wind power, in addition to ocean currents, is one of the useful energy resources in NTT. A wind power plant (Pembangkit Listrik Tenaga Bayu, PLTB) is one alternative to be suitably developed in the coastal areas. The contour map of wind speed (Fig. 2.3) marks out different wind speed between regions in Indonesia. Figure 2.3 shows that NTT is situated on the periphery of Indonesia with wind speeds ranging between 4.75 m/s (meters per second) to 5.5 m/s. The wind speed in NTT is higher than in Kalimantan, reaching to a maximum rate of 2.5 m/s. This is because NTT lies in the southernmost part of Indonesia and, thereby, is closely exposed to strong winds in the unblocked high seas. Meanwhile, Table 2.6 described several locations in NTT with the potential of wind speed (v) and power (W) per square meter harnessed from the surface area in windmill blades.

However, the development of PLTB in NTT is still very limited. In 2013 PLN generated 90 kWh of wind power in NTT, far less than its potential (PLN, 2013b). The issues of technology, costs, and stability are among the barriers for an optimum utilization of wind resource in the province. In 2018, PT PLN started to build the

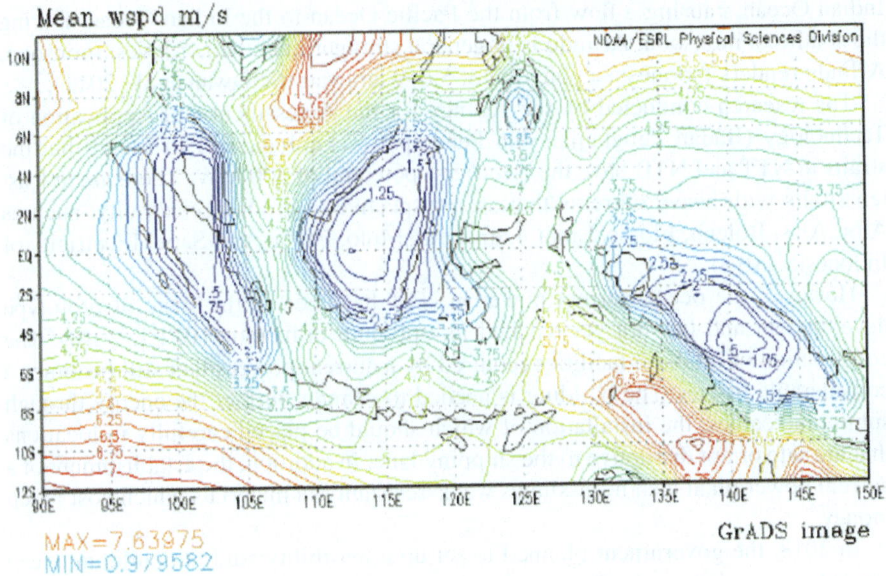

Fig. 2.3 Contour map of wind in Indonesia. *Note* Ion: plotted from 90.00 to 150. Iat: plotted from −12 to 11.00. t: averaged over Jan 1969 to Dec 1978. lev: 0. *Source* Susandi (2012)

first PLTB in East Nusa Tenggara with a capacity of 20 Megawatts (Mw). The PLTB, which was set up to be built in Oelbubuk Village, Molo Tengah District, South Central Timor, according to the 2019–2028 Electricity Supply Business Plan (RUPTL), will be funded by Independent Power Producer (IPP) funding sources. Based on the data from the NTT Energy and Mineral Resources (ESDM) Office, the potential for PLTB in the area reaches 10,188 MW, but the installed capacity has only reached 3.1 MW or 0.03% (Amalo, 2020).

2.3.5 *Solar Power Plant*

Solar resource is considerably potential in NTT thanks to the duration of irradiation which is around eight months with an estimated daily intensity of 5.1 kWh/m^2. This great potentiality will enable the province to cover all its electrical energy needs (Likadja, 2014). Solar power is useful to raise the capacity of diesel-powered generators during peak load demands. The NTT government itself has planned to build solar power plants, and the expansion targeted 14 districts or cities on the island. In 2016, it erect PLTS of 1 × 400 kW in Southwest Sumba District, PLTS of 1 × 300 kW in East Sumba District, PLTS of 1 × 75 kW and PLTS of 1 × 50 kW in Rote Ndao District, PLTS of 1 × 75 kW in West Manggarai District, and PLTS with a total capacity of 330 kW in Alor District (Ministry of Energy and

Table 2.6 Wind speed in NTT

No.	District	Sub-District	Location	v (m/s)	Power (W/m^2)
1	KUPANG	Semau	Hansisi	4.20	259,034
		Kupang	El tari	3.13	107,509
		Tengah	Lasiana	2.83	79,232
		East Kupang	Oesao	3.10	104,158
		East Sabu	Tordamu	3.78	188,909
2	TTS	West Amanuban	Nusa	4.30	27,798
		South Molo	Fatukolen	7.00	
3	Rote Ndao	Lobalain	Lekunik	2.75	72,691
		Central Rote	Maubesi	4.10	240,968
4	Flotim	Larantuka	Larantuka	1.63	0,000
5	Sikka	Paga	Masabewa	3.10	104,158
		Maumere	Maumere	2.21	37,565
6	Ngada	Aimere	Sebo Woli	3.20	114,567
7	Manggarai	Komodo	TN.Komodo	3.00	94,400
		Ruteng	Satar Racik	3.88	204,222
		Lembor	Nanga Lili	4.50	318,600
		Borong	Nanga Lebang	3.30	125,646
		Satarmeze	Papang	2.80	76,751
8	Lembata	Nagawutung	Pasir Putih	3.50	149,904
9	Sumba Timur	Kota Waingapu	Mau Hau	1.73	0,0000
		Haharu	Mondu	4.60	340,315
10	Sumba Barat	Peberiwai	Kamanggih	4.01	225,445

Source Likadja (2014)

Mineral Resources, 2016). The advancement of solar power plants is an alternative to accelerate the electrification ratio. The NTT Mining Service has run a solar power generation program integrated with Solar Home System (SHS). In 2013, the NTT Provincial Government built approximately 40 units of SHS for nine villages under the scheme of the Special Allocation Funds (DAK) managed by the district or city local government.

There are a total of 7.5 MW of PLTS spread across the NTT region. One of the PLTS also been considered as the largest PLTS in Indonesia, which is located in Bajaneke Hamlet, Oelpuah Village, Kupang, NTT. With a capacity of 5 MW, the plant owned by the National Electronic Institute (LEN) helps PLN overcome the deficit in the Timor system, where rotating blackouts during the day can be resolved since the end of December 2016, the beginning of the operation of this PLTS. Twenty-two thousand of solar modules stretching over an area of 7.5 ha, with one module producing about 230 watts of electricity, which accounts for 4% of the total electricity needs of Timor Island. This energy is then sold to PLN at a price of US $ 25 cents per kWh with a 20-year contract period, which will be distributed to consumers as needed (Kementerian ESDM RI, 2020; Simorangkir, 2016).

Currently, the central government will also build PLTS in the tourism Special Economic Zone (KEK) of Labuan Bajo, NTT. The first phase will be constructed in Tana Mori and Tana Naga in early 2021 at the latest and will have a capacity of 70 MW from a total of 210 MW (Indrawan, 2020). The government will also set up communal PLTS in 11 areas that have not yet electrified or have not connected by the on-grid electricity (Budiartie, 2019).

2.3.6 *Geothermal Power Plant*

Geothermal energy has some advantages compared to conventional ones, such as petroleum and coal. Although the drilling is very costly and risky, geothermal energy brings some benefits, i.e. (1) low carbon emissions; (2) lack of weather constraints; (3) sustainability (4) small-to-large scales; and (5) void of radiation (Setiawan, 2012).

NTT province, especially Flores island has considerable geothermal potential, including in Ulumbu, Mataloko, Mutubusa, Ropa and Atadei. The potential of the geothermal resource in NTT accounts for roughly 1323.5 MW spreading in 24 areas in 9 districts in NTT (Ministry of Energy and Mineral Resources, 2015). Provided that this potential is maximized, PLTP alone would be enough to cover all the electricity needs on the island of Flores (Likadja, 2012).

In the Book of Indonesia's Geothermal Potential Volume II in 2017, it is stated that of the various geothermal potential site in NTT, six of them have been designated as WKPs, and of the six WKPs have entered the exploitation stage. The following are the geothermal potential in NTT. *First*, WKP Atadai, which located in Lembata Regency, is held by PLN and currently is still in the exploration stage. From the estimated potential of 40 MW, it is planned that 10 MW will be developed in 2020. *Second,* WKP Gunung Sirung, located in Alor Regency, has the potential to suspect 152 MW, is in preparation for an auction or assignment. *Third,* WKP Mataloko, located in Ngada Regency, has a PLTP consists of a hypothetical potential of 10 MW, an estimated 6.5 MW and a current installed capacity of 2.5 MW. The Matoloko geothermal exploitation permit is held by PT PLN (Persero). *Fourth,* WKP Oka-Ille Ange is located in East Flores Regency and has an estimated potential of 50 MW with a 10 MW development plan in 2024. *Fifth,* WKP Sokoria, located in Ende Regency, is in the exploration stage with the concession license held by PT Sokoria Geothermal Indonesia. Sokoria's geothermal potential are scattered in several places, namely Komandaru (11 MW), Ndetusoko (10 MW), Sokoria with a hypothetical potential of 145 MW and an estimated 25 MW, then Jopu (5 MW). The Sokoria WKP was developed in several stages. Phase I is 5 MW in 2019, Phase II in 2019 is 5 MW, Phase III is 10 MW in 2020, and Phase IV is 10 MW in 2021. *Sixth,* WKP Ulumbu, which located in Manggarai Regency, has an estimated potential of 100 MW and proved to be 12.5 MW (Directorate of Geothermal, 2017).

2.3.7 *Bioenergy Power Plant*

Bioenergy is energy made available from organic materials. It has various form of solid (biomass), liquid (biofuel), or gas (biogas). Biomass can be produced from plant products or wastes. There are numerous types of plants commonly used for biofuels, such as oil palm, cassava and sugar cane. Husk, straw, and *merang* are examples of plant wastes usually used for biofuels. The popular types of biofuels are bioethanol, biodiesel, and pure plant oil, as a substitute for respectively premium, diesel fuel, and kerosene or diesel fuel. Animal manures, such as those from cows, are useable to produce biogas.

The bioenergy potential in NTT has not yet been utilized. The cause is, for instance, a relatively little rainfall that prevents the plants in NTT from growing well like in Sumatra and Java. Another reason is that most of the livestock systems in the region are spatially scattered. Most of the farmers just release their livestock anywhere to find foods, making it difficult to harvest their manure for biogas.

Furthermore, from all potential renewable energy resources in NTT, the most feasible to be developed as a source of electricity are solar, geothermal, and hydro, due to its large potential and economically feasible. The government has also boosted the use of renewable energy through a massive expansion of PLTS as well as constructing more PLTP and PLTMh throughout the island. One of the districts that is suitable for the development of these renewable energies is Manggarai Regency. Manggarai also has several tourist spots and economic areas which indeed requires a large electricity supply. Thus, the following sub-chapters will discuss the potential energy, the electrical demands and their provision in order to ensure sustainable economic development in Manggarai District.

2.4 Condition of Power Plants in Manggarai District

Manggarai District is a parent district which has undergone two regional proliferation. It encompasses the plains of Flores Island and Mules Island, with the land area of 1915.62 km^2. Manggarai District consists of 11 sub-districts bordering with the Flores Sea in the north, Sawu Sea in the south, West Manggarai District in the west, and East Manggarai District in the east. The dominant feature of Manggarai District is a plateau with a slope of more than 40° with a total area of 89,603 ha (53.67%).

Manggarai District has a quite diverse energy potential. The prospects of new and renewable energy belong to geothermal, water, solar and wind, available throughout the Manggarai region. One potential that has been cultivated by the central and regional governments is PLTP Ulumbu in Satarmese Sub-District, with a power capacity at 4 × 2.5 MW. There are also two PLTMh that have been operating, i.e. Wae Garit in Langke Rembong Sub-District and Hilihintir in West Satarmese Sub-District. The regional government also decided to build PLTMh Wae Racang and PLTMh Wae Lega in North Rahong Sub-District (Bappeda, Manggarai District, 2014).

Besides geothermal and hydro, Manggarai District also makes use of solar energy for power generation. Currently, a number of Solar Power Plants (PLTS) has been distributed in several sub-districts of Manggarai District, which are Cibal (Golowoi Village, at 2.75 kW and Riung Village, at 1.3 kW), Wae Ri'I (Ndehes Village, at 1.25 Kw and Wae Mulu Village, at 1.30 kW). Then, there are PLTS in other sub-districts which are Lelak (Urang Village, at 5.25 kW; Bangka Lelak Village, at 1.30 kW; and Gelong Village, at 1.25 kW), and North Rahong (Benteng Village, at 1.3 kW). Moreover, PLTS are also located in West Satarmese (Gulung Village, at 0.85 kW, Nuca Molas Village, at 1.5 kW, and Hilihintir Village, at 1.6 kW), and West Reok (Sambi Village, at 20 kW, where there are also centralized PLTS and SHS) (Bappeda, Manggarai District, 2014).

The development of new and renewable energy needs to connect with people's economic activities. It is indisputable that the economic condition will determine a supply–demand rate for new and renewable energy. In 2010–2012, the economy of Manggarai District grew at a range of 5.3–5.6% per annum (Bappeda, Manggarai District, 2014). Given the positive growth rate, the people's per capita income continued to increase from the US $ 4003 in 2010 to the US $ 4793 in 2012 (Bappeda, Manggarai District, 2014). Although the economic growth was quite promising, and the per capita income kept rising, the poverty rate was still so high. There were 42,845 households living below the poverty line in nine sub-districts in Manggarai District. The highest was in Ruteng with 5679 low-income families, followed by West Satarmese, Langke Rembong, and Satarmese (the research site) with respectively 4830, 4635, and 4624 households. Lelak Sub-District, which is bordering with West Manggarai District, accounted for the lowest poverty rate with 1940 households. Good access roads in these sub-districts correlated with the relatively low number of poor households (Bappeda, Manggarai District, 2014).

Agricultural sector dominates the economic activities in Manggarai District. In 2012, this sector had a relatively low growth at 1.1%. The mining and quarrying sectors also underwent a decline in growth rates, from 3.6% in 2011 to 1.7% in 2012. The sectors of trade, hotel and restaurant also moved downward from 10.5% in 2011 to 8.1% in 2012. In spite of the economic slowdown in a number of sectors, the 2013 Manggarai District Economic Indicators showed that the service sectors, including financial, leasing and business, experienced significant growth of 10.6% in 2012.

In fact, the utilization of new and renewable energy in Manggarai District has been carried out for quite a long time. The development of new and renewable energy in Manggarai is parallel to its potential and natural conditions. The highland contour of the region inspires the use of PLTMh for a source of power generation. As the guard officer said, PLTMh Wae Garit was built since 1978, and Ulumbu geothermal was developed since 1996 by Genzl New Zealand (PLN, 2014).

Prior to the operation of the new units of PLTP Ulumbu, Manggarai District underwent an electricity deficit. The data from PLN showed that the peak load was recorded at 7.4 MW in February 2014. However, the electricity supply in the same month was only 7.2 MW for which PLTP Ulumbu contributed 3.8 MW, while PLTD

Waso and PLTMh covered the rest (PLN, 2014). It means that under this condition, the district had a power shortage of 0.2 MW. This deficit was resolved since the operation of PLTP Ulumbu Unit 3 and Unit 4 with a capacity of 2 × 2.5 MW, reducing at the same time the operational hours of PLTD Waso. As a result, the local PLN has successfully saved Rp. 75 billion per year on diesel fuel subsidies. This is what our informant from PLN Ruteng said during an interview in 2014.

Thanks to the development of geothermal resources, Manggarai District has now reached an electricity surplus. The irony is that not all the Manggarai communities benefit from this power surplus given that some countryside areas have not yet electricity access due to lack of grid and low demand for electricity. The electricity demand in this region, nevertheless, will continue to grow along with the improvement of strategic economic areas and tourism, as well as smelters planned to build in the southern region of Manggarai. Therefore, the regional government should keep providing electricity infrastructures on the basis of new and renewable energy to support regional economic activities.

2.4.1 Energy Potentials and Condition in Satarmese

Satarmese, where the research was conducted, is situated in Wewo Village (Tantong Hamlet and Damu Hamlet) and Lungar Village (Lungar Hamlet and Mesir Hamlet). It is one of the sub-districts in the administrative area and in the southern part of Manggarai District. The southern part of Satarmese Sub-district is directly adjacent to the Savu Sea, the northern to Ruteng and Langke Rembong Sub-districts, the western to West Satarmese Sub-district, and the eastern to East Manggarai District. Satarmese has an area of 298.49 km^2 with a population rate at 33,870 people and a population density at 113.47 people/km^2 in 2012 (BPS, 2013).

This sub-district, with Iteng being as the capital, is bestowed with a high diversity of flora and fauna. The most common plantation crops are coffee, cloves, vanilla, coconut, and candlenut. Geographically, the topography of Satarmaese is dominated by slopes or hills. The average height of this region is 530 meters above sea level (BPS, 2013). Geographically composed of hills mostly on the edge of a protected forest area, this sub-district has a great potential for the development of new and renewable energy resources of geothermal and water in support of PLTP Ulumbu and PLTA.

Satarmese Sub-district has also a great potential of energy resources, such as geothermal, solar, air, and wind. The geographically hilly area renders to this region a very promising hydro resource from upstream springs. A potential wind resource is found in the southern coastal region, while solar in all over regions of Satarmase. However, poor access to the location makes it hard to exploit these natural resources.

Some communities in Wewo and Lungar villages, which were the focus of this research, have not yet gained access to electricity. This is an irony because they are relatively close to the Ulumbu geothermal well. These areas failed to fully utilize the geothermal resource (PLTP-Ulumbu). Wewo Village has eventually gained access to electricity networks since March 2014, while Lungar Village still has to wait longer. The following sub-chapters will discuss to what extent the two villages make the best use of their energy resources and potentials.

2.5 Energy Potentials and Utilization in Wewo Village

Wewo Village is one of the areas that became the focus of this research. The village is located at an altitude of 536 m above sea level. With the contour of hilly areas, it also has a considerable potential for renewable energy consisting of geothermal, water and solar. It is unfortunate that these three potential alternative energy sources have not been fully utilized. Now, only geothermal and solar have been cultivated for lighting.

Despite its many energy potentials, not all hamlets in Wewo Village benefit from these abundant resources. Water and geothermal sources (see Fig. 2.4) are very close to Tantong Hamlet. In fact, the upstream springs running through Tantong Hamlet could be used for PLTMh rather than, as it is now, just for a cooling water for PLTP. Sadly, Tantong Hamlet, which was also the main locus of this research, did not receive an advantage from the nearby energy resources. In March 2014, the hamlet finally got access to geothermal-powered electricity. Earlier on, the people of Tantong Hamlet relied on lamps, torches, or flashlights for daily lighting. Some households got access to electricity through electrical cables connected to the nearest hamlet, i.e. Damu Hamlet. PLTP Ulumbu is only about half a kilometer away from the nearest

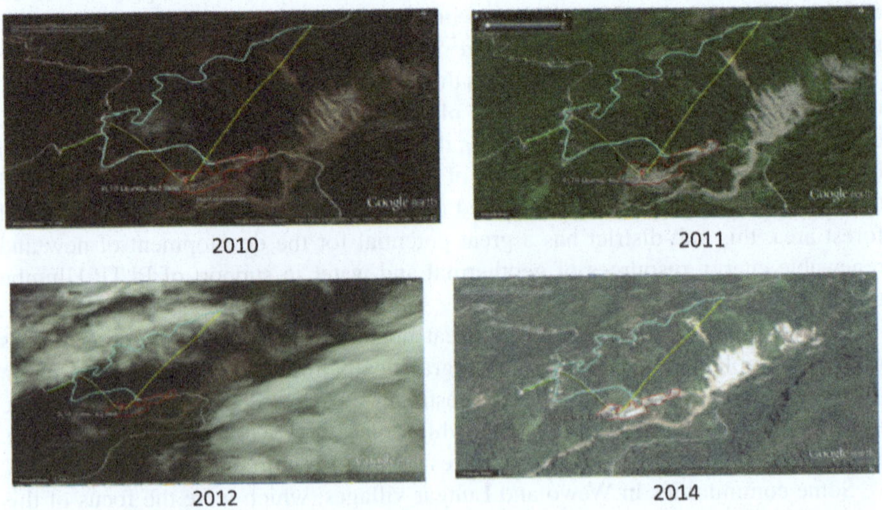

Fig. 2.4 Location of PLTP Ulumbu and Tantong Hamlet. *Source Google Earth*, processed

house (*Google Earth*, 2014), and less than two kilometer from the furthest house in Tantong Hamlet (*Google Earth*, 2014).

The people of hamlet has gained access to electricity since 2003 long before the operation of PLTP Ulumbu. However, not all households enjoyed electricity access in that year. The Chief of the Wewo Village said that this was because of the high cost of electricity installation amounting to seven million rupiah for each house. Nevertheless, the PLN electricity installation rate continued to raise at an additional 34 families in 2014.

The current utilization and development of geothermal resource have not been optimal in supporting the regional economic activities. At present, the installed capacity of PLTP Ulumbu is just 10 MW. The field visit found that the electricity supply from the PLTP was only about 2×2.5 MW by the end of 2014, mostly coming from Unit 3 and Unit 4 because Unit 1 and Unit 2 stopped to operate for maintenance.

The water resource has been so far utilized in a very limited way, even though its potential in Wewo Village is reasonably promising to support the development of new and renewable energy. The somewhat heavy water of the big river Wae Ces is passing through the village. However, the field research found only about 12 out of 98 households in Tantong Hamlet and Damu Hamlet have been making a practical use of hydropower (PLTMh or Mini-Hydro Power Plant) for lighting. The contour of hilly areas has a large potential for the development of water-powered electricity in this region. The potential water resource is very useful to supply the cooling tower of PLTP Ulumbu. Figure 2.5 presents the profile of land elevation in Tantong Hamlet from the upstream part of the river towards the PLTP Ulumbu. The research site of Tantong Hamlet is situated at an altitude of up to 883 m above sea level.

Tantong Hamlet and Wewo Village are rich of potential water resource. Thus, the presence of PLTMh in Wewo Village is expected to bring energy independence and economic improvement to the surrounding community, while keep watering the PLTP Ulumbu and the people's rice fields.

Wewo Village has a lesser potential energy of Sun's rays and wind than geothermal or water resource. A ray of sunlight comes through the village only four to six hours a day due to the contour of the area which is mostly hilly. Nevertheless, Solar Home System (SHS) or SEHEN (PLN) has been installed in some of the neighboring villages under a similar condition. The SEHEN is capable of turning on the lights

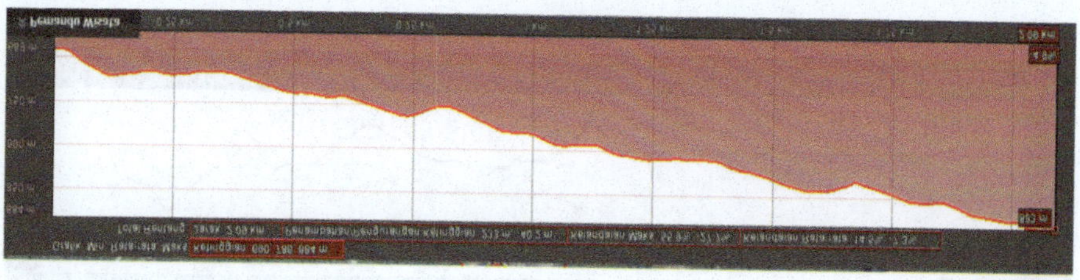

Fig. 2.5 Profile of land elevation from Tantong Hamlet (top right) to PLTP Ulumbu (left corner). *Source Google Earth*, processed

with a total capacity of around 12 Watts, while the SHS given for free by the ESDM Ministry has a higher capacity or at least 60 W. The sustainability of solar power utilization, especially SHS, is still questionable. Complicated technologies and lack of sufficient spare part maintenance and replacement were among the barriers. As for SEHEN, the community was having a difficulty to pay the monthly electricity bills because of the long distance between the houses and the PLN payment point and the relatively expensive transportation costs. The 2013 Electricity Team reported that the rural electricity development has to deal with an overlap between the programs run by different agencies. Unless this condition is immediately resolved, it will hamper the advancement of solar energy in the long run.

2.6 Energy Potentials and Condition in Lungar Village

The research site of Lungar Village is located at an altitude of 1253 m above sea level (Fig. 2.6). The contour of the area is not much different from Wewo Village. As seen in Fig. 2.6, the contour of the village is hilly with an extremely varied land elevation which at most can reach 1720 m above sea level. Interviews and field observations showed that Lungar Village has the same energy potentials as Wewo Village, i.e. solar, water, and geothermal resources. However, some challenges exist in Lungar Village making it somewhat slower in the utilization and development of its energy resources. First, Lungar Village has more difficult access roads than the neighboring Wewo Village. Second, the households in Lungar Village have a fairly considerable distance from each other. As a result, building infrastructures for electricity networks in the area is relatively expensive.

The development of solar resource in Lungar Village is one step ahead the utilization of geothermal and water energy. PLN has installed SEHEN in this village giving service to about 96% of 128 households. The SEHEN program, which belongs to PLN, helped improve people's access to lighting. However, this program was not without problems. First, the electricity capacity or power was relatively low. Second, the electrical power was mainly reliant on the intensity of solar exposure. The topography of Lungar Village, which is hilly and often foggy, leads to a relatively low solar radiation. Third, the program was not free of charge for which PLN has imposed a monthly maintenance fee. To pay their electrical bills, some local residents claimed

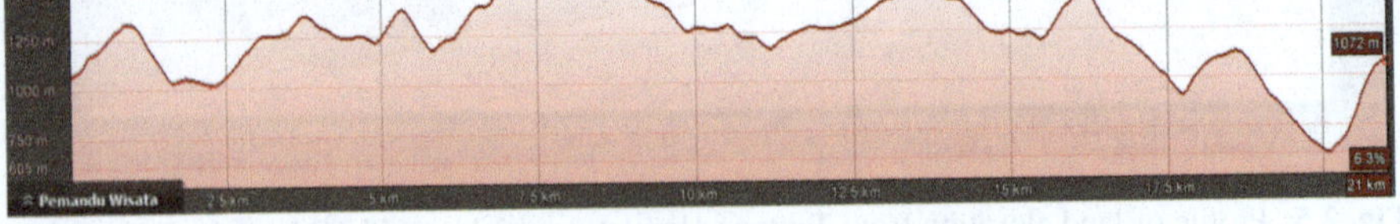

Fig. 2.6 Profile of land elevation in Lungar Village. *Source Google Earth*, processed

that they have to go to Ruteng, which is quite far away and, thereby, costing them much money for transportation. As an illustration, in order to get to Ruteng from Lungar, they had to spend around Rp. 70 thousand for a round trip travel, which is indeed higher than the electricity contribution of Rp. 30 thousand per month. The Head of Lungar Village said that the residents must spend at least Rp. 100,000 for the SEHEN monthly contribution. This also implies that the cost for the SEHEN payment was also relatively high. All these obstacles have put the sustainability of the SEHEN program into question.

Water energy is an alternative new and renewable energy that can benefit the people of Lungar Village. There is a waterfall in Mesir Hamlet potentially useful for micro hydro power plants. In addition, Lungar Village also has a potential geothermal resource. The distance between Lungar Village and Poco Leok Volcano is about three kilometers with an elevation or a slope angle at 45° where the geothermal energy potential is situated. The hilly contour and proximity to the volcano are an indication of potential geothermal energy and, thereby, is worth considering. Moreover, the locations of PLTP Ulumbu, PLTP Darajat (Garut), and PLTP Kamojang PLTP (Garut), are nearby the volcano. Figure 2.7 shows the distance between Poco Leok Volcano, Lungar Village, and PLTP Ulumbu.

It is evident from the satellite image in Fig. 2.7 that Lungar Village has a considerable geothermal potential. If an elevation of 650 m provides the PLTP Ulumbu with the capacity of 100 MW, an altitude of 1022 m should have granted Lungar Village with a great potential of geothermal resource. Sadly, the geothermal resource in this region has not been identified accurately due to lack of exploration. Access to the region is also difficult due to a curving and steep geographical condition. Furthermore, the electricity demands in the region are fairly low. An informant from PLN Ruteng stated that a relatively large electricity demand came from the touristic places in Labuhan Bajo and the food processing industry. Thus, the development and utilization of geothermal and water resources in this region are in need of serious attentions and thoughts.

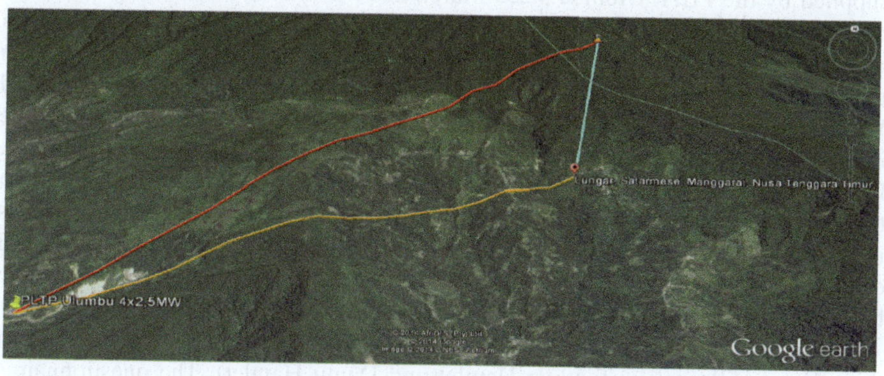

Fig. 2.7 Satellite image of Volcano Poco Leok, Lungar Village, and PLTP Ulumbu. *Source* Google Earth, processed

2.7 Profile of Electricity Demand and Supply in the Research Sites

Energy utilization in a given area is inseparable from its energy and economic potentials. The need for energy is related to the size of regional economic activities. The availability of supporting facilities and infrastructures, and the intensity of economic activities in any region are contingent upon the energy supply. These two factors are decisive in improving the economy of rural communities. The following section is an attempt to look at the conditions of these two factors in Wewo Village and Lungar Village which were the focus of the research.

The questionnaires showed that the expenditure in Wewo Village touched an average level of Rp. 643 thousand per month in 2013 and Rp. 706 thousand per month in 2014. The rates were still below the Regional Minimum Wage (UMR) which is one million rupiah per month (BPS, 2013). According to the survey in 2014, people in Damu Hamlet and Tantong Hamlet spent their money mostly on foods and beverages, energy (electricity, gas, kerosene, etc.), ritual meals (*kenduri*) and education. The monthly expenditure on foods and beverages was Rp. 153 thousand (21.7%); on energy Rp. 128 thousand (18.1%); on ritual meals Rp. 109 thousand (15.5%); and on education Rp. 96 thousand (13.6% or the lowest). It can be inferred from the data that the people's need for energy ranked the second largest of the total expenditure. Some people were still relying on kerosene and fuel wood for energy sources.

The same survey data also implied that the demand for energy, especially electricity, has a significant proportion to the total household expenditure. It is problematic that while the people's need for electricity will most likely continue to increase in Tantong Hamlet and Damu Hamlet, the purchasing power of the residents is rather low. The slow progress of industrial and productive activities became a major hindrance to the improvement of energy demands.

The people of Tantong Hamlet and Damu Hamlet have been enjoying electricity supplied by the PLTP Ulumbu since March 2014 and 2003 respectively. PLN has been also delivering the electricity surplus from the PLTP Ulumbu to a number of areas in Manggarai District. Currently, the electricity supply from the PLTP has been mostly used to meet the electricity needs outside Tantong Hamlet in particular and Wewo Village in general. The central and regional governments need to notice some important things in regard with the location of Tantong Hamlet within Wewo Village. Economic incentives and community empowerment are high of importance to sustain the improvement of economic and productive activities among the local communities in Wewo Village, especially in Tantong Hamlet. Although Damu Hamlet has gained electricity supply since 2003, the field observation proved that there is no productive business or a home industry being developed.

Lungar Village is located at the forest's edge and has a better economic condition than Wewo Village (i.e. Tantong Hamlet and Damu Hamlet). The questionnaires found that the average household expenditure in Lungar Village is Rp1.3 million per month. In spite of a higher level of expenditure, Lungar Village has been denied access to the PLN electricity grids. Nowadays, the inhabitants of Lungar Village depend on the PLN's SEHEN electricity program limited to lighting only. Although their economic condition is relatively better, a poor access road has delayed the installation of the PLN electricity networks until June 2014.

According to the 2014 survey data, the people of Lungar Village spent most of their money in three major areas, i.e. ritual meals (*kenduri*), foods and beverages, and education. The average expenditure for ritual meals was Rp. 616 thousand per month (46.8%). The second largest spending was on foods and beverages, which is Rp. 278 thousand per month (21.1%); education followed with an average expenditure of Rp. 163 thousand per month (12.4%). The energy demand, especially for lighting, in the region was in fact more than enough to support ritual meals and other activities at night. However, the electricity supply in Lungar Village was still confined to solar resources (or SEHEN) and generator sets.

The condition of Lungar Village, which entirely relies on solar energy and kerosene for lighting, is a quite contrast to that of Wewo Village where most people already have access to electricity. The power supply from PLTP Ulumbu should have been able to flow electricity to the furthest possible in Lungar Village. The main problem is the absence of electricity networks distributed to the households in the area, mostly because of a difficult geographical condition. The presence of the PLN electricity networks in the area will save diesel fuel subsidies allocated so far mainly for running generator sets. It is imperative for PLN and the regional government to respond as quickly as possible the need for electricity networks in Lungar Village. In the future, other alternative ways of electricity access improvement need to be explored in order to meet the people's growing need for electricity.

2.8 Conclusion

In conclusion, NTT Province, as well as the research areas, have a relatively diverse potential of primary energy resources, such as geothermal, solar, wind, hydro and ocean currents. Under good management, these primary energy resources will answer the energy needs in the region. However, this study found that the potential of renewable energy in NTT has not reached its optimum utilization due to various limitations, such as unstable energy supplies, high investment start-up costs, natural constraints, and lack of appropriate technology application. The underdeveloped new and renewable energy resources indicate that the development of this sector is urgently required.

References

Amalo, P. (2020, September 6). The first Bayu power plant will be built in NTT. *Pembangkit Listrik Tenaga Bayu Pertama akan Dibangun di NTT.* Retrieved September 6, 2020, from https://mediaindonesia.com/read/detail/274972-pembangkit-listrik-tenaga-bayu-pertama-akan-dibangun-di-ntt

ANTARA News. (2019, December 10). PLN NTT is targeting to build geothermal power plants of up to 115 MW. *PLN NTT targetkan bangun pembangkit panas bumi hingga 115 MW.* Retrieved September 5, 2020, from https://www.antaranews.com/berita/1202135/pln-ntt-targetkan-bangun-pembangkit-panas-bumi-hingga-115-mw

Bappeda, Manggarai District. (2014). The general information about electricity development in Manggarai District. *Presentasi mengenai gambaran umum pembangunan kelistrikan di Kab. Manggarai*, June 2014.

BPPT. (2010). Ocean wave power plant Prototipe Trial. *Uji Coba prototipe pembangkit listrik tenaga arus laut.* Jakarta: BPPT.

BPS. (2013). Satarmese district in 2013 figures. *Kecamatan Satarmese dalam angka 2013.* BPS of Manggarai District, East Nusa Tenggara.

BPS of East Nusa Tenggara. (2014). East Nusa Tenggara in 2014 figures. *Nusa Tenggara Timur dalam angka 2014.*

BPS. (2019). East Nusa Tenggara province in figures 2019. *Provinsi Nusa Tenggara Timur Dalam Angka 2019.*

BPS Provinsi NTT. (2018a). Total Population by age group and regency/City in East Nusa Tenggara Province, 2017. *Jumlah Penduduk Menurut Kelompok Umur dan Kabupaten/Kota di Provinsi Nusa Tenggara Timur,* 2017. Retrieved September 5, 2020, from https://ntt.bps.go.id/dynamictable/2020/06/25/1481/jumlah-penduduk-menurut-kelompok-umur-dan-kabupaten-kota-di-provinsi-nusa-tenggara-timur-2017.html

BPS Provinsi NTT. (2018b, January 30). Projected population growth rate of East Nusa Tenggara Province. *Proyeksi Laju Pertumbuhan Penduduk Provinsi Nusa Tenggara Timur*, 2011–2020. Retrieved September 5, 2020, from https://ntt.bps.go.id/dynamictable/2016/11/23/361/proyeksi-laju-pertumbuhan-penduduk-provinsi-nusa-tenggara-timur-2011-2020.html

Budiartie, G. (2019, April 12). This Eastern heaven of Indonesia will be bright with solar power. *Surga Timur RI Ini Bakal Terang dengan Tenaga Surya.* Retrieved September 7, 2020, from https://www.cnbcindonesia.com/news/20190412142441-4-66328/surga-timur-ri-ini-bakal-terang-dengan-tenaga-surya

Directorate of Geothermal, Directorate General of EBTKE, Center for Mineral, Coal and Geothermal, B. G. (2017). Indonesia's Geothermal Potential Volume 2. *Potensi Panas Bumi Indonesia Jilid 2.* Jakarta.

Government Province of NTT. (2010). Management of coal and mineral mining. *Pengelolaan pertambangan batu bara dan mineral. Perda Provinsi NTT No. 8 Tahun 2010 tentang Pengelolaan Pertambangan Batu Bara dan Mineral.*

Indrawan, R. (2020, March 9). PLTS Labuan Bajo 70 MW Phase I is targeted to be completed in 2022. *PLTS Labuan Bajo 70 MW Tahap I Ditargetkan Rampung 2022.* Retrieved September 7, 2020, from https://www.dunia-energi.com/plts-labuan-bajo-70-mw-tahap-i-ditargetkan-rampung-2022/

Januwarsono, S. (2010). Electricity system development planning as an effort to meet the needs of electric energy in the province of NTT. *Perencanaan pembangunan sistem kelistrikan sebagai upaya pemenuhan kebutuhan energi listrik di provinsi NTT.*

Lewokeda, A. (2019, February 21). The use of electrical energy in NTT grew by around 8.4%. *Pemakaian energi listrik di NTT tumbuh sekitar 8,4 persen. ANTARA News Kupang, Nusa Tenggara Timur.* Retrieved from https://kupang.antaranews.com/berita/14139/pemakaian-energi-listrik-di-ntt-tumbuh-sekitar-84-persen

Likadja, F. J. (2012). Analysis of the implementation of the national energy policy on energy demand in the province of East Nusa Tenggara using leap software. *Analisis penerapan kebijakan energi nasional terhadap permintaan energi di provinsi Nusa tenggara timur menggunakan perangkat lunak leap. Prosiding Seminar Proceeding Nasional Sains dan Teknologi, Vol. 2 Bidang Teknik, T-181–T-187.* Bali: Undana Press.

Likadja, F. J. (2014). Implementation of conservation and diversification policies on the availability of electrical energy in the province of East Nusa Tenggara. *Penerapan kebijakan konservasi dan diversifikasi terhadap ketersediaan energi listrik di provinsi Nusa Tenggara Timur*. Nusa Cendana Kupang University. Retrieved October 15, 2014 from http://ipenconsulting.com/yahoo_site_admin/assets/docs/P15_FLikadja_Re-view_of_RE_Policy_in_NTT.4825204.pdf

Ministry of ESDM. (2015). 2015 EBTKE Statistics. *Statistik EBTKE 2015*. Jakarta: Ministry of ESDM, Directorate General of New, Renewable Energy, and Energy Conservation.

Ministry of ESDM. (2016). Development of new renewable energy-based power plants in East Nusa Tenggara Province, Fiscal Year. Pembangunan Pembangkit Listrik Berbasis Energi Baru Terbarukan di Provinsi Nusa Tenggara Timur Tahun Anggaran 2016–2017. Jakarta: Ministry of ESDM, Directorate General of New, Renewable Energy, and Energy Conservation.

Ministry of ESDM RI. (2018, April 2). The minister of energy and mineral resources visits the location of the construction of a sea current power plant in the Larantuka Strait. *Menteri ESDM Tinjau Lokasi Pembangunan Pembangkit Listrik Arus Laut di Selat Larantuka*. Retrieved September 5, 2020, from http://ebtke.esdm.go.id/post/2018/04/02/1924/menteri.esdm.tinjau.lokasi.pembangunan.pembangkit.listrik.arus.laut.di.selat.larantuka

Ministry of ESDM RI. (2020, February 17). The government continues to improve PLTS development. *Pemerintah Terus Perbaiki Tata Kelola Pembangunan PLTS*. Retrieved September 7, 2020, from http://ebtke.esdm.go.id/post/2020/02/17/2476/pemerintah.terus.perbaiki.tata.kelola.pembangunan.plts

PLN Sub-rayon Ruteng. (2014). 2013 realization in generating sector. *Realisasi 2013 bidang pembangkit*. Jakarta: PT PLN.

PLN. (2013a). Electricity supply business plan. *Rencana usaha penyediaan tenaga listrik PT PLN 2013–2022*. Jakarta: PLN.

PLN. (2013b). PLN Statistics 2013. *Statistik PLN 2013*. Jakarta: PLN.

PLN. (2014a). Historical development of the Ulumbu geothermal field. *Historical pengembangan lapangan panas bumi Ulumbu*. Jakarta: PT PLN.

PLN. (2014b). Determination of the tariff adjustment for electricity (tariff adjustment) in November 2014. *Penetapan penyesuaian tarif tenaga listrik (tariff adjustment) bulan November 2014*. Accessed on November 5, 2014 from http://www.pln.co.id/dataweb/TTL2014/07_Tariff%20Adjusment%20Nopem-ber%202014.jpg

Republic of Indonesia. (2007). Electricity. Indonesian Law No. 30/2009 about Electricity. Ketenagalistrikan. *Undang-undang No. 30 tahun 2009 tentang ketenagalistrikan*

Setiawan, S. (2012). Geothermal energy within the MP3EI framework: Analysis of prospects, constraints and policy support. *Energi panas bumi dalam kerangka MP3EI: Analisis terhadap prospek, kendala, dan dukungan kebijakan. Jurnal Ekonomi dan Pembangunan, XX* (1).

Setiawan, F., Hernandhy, E., & Qisthy, A. (2011). Renewable energy development by identifying cross-Indonesian flows in Eastern Indonesia. *Pengembangan energi terba- rukan dengan identifikasi arus lintas Indonesia di Wilayah Timur Indonesia*. Bandung: Unpad.

Simorangkir, E. (2016, September 22). Take a peek at the Solar Power Plant. *Mengintip Pembangkit Listrik Tenaga Surya Terbesar di RI*. Retrieved September 6, 2020, from https://finance.detik.com/energi/d-3351258/mengintip-pembangkit-listrik-tenaga-surya-terbesar-di-ri

Susandi, A. (2012). Wind data processing and wind energy potential in Indonesia. *Pengolahan data angin dan potensi energi angin di Indonesia*. Retrieved November 4, 2014, from www.armisusandi.com/energi/*P5_ME4132.ppt*.

Chapter 3
Political Economy of Geothermal Utilization: Cultivating the Potential of Ulumbu for Welfare Improvement

3.1 Introduction

The lack of electricity access in Eastern Indonesia is one of the main problems in this country. The case of Manggarai District, which is about three hours driving to the east of Labuan Bajo, is clear evidence. The electrification ratio in Manggarai District in 2012 was only at 36%, far below the national average of above 76% (Mining and Energy Service of Manggarai District, 2014). This situation is so unfortunate given that the area has a considerable energy potential that can be further developed.

One of the key energy resources in East Nusa Tenggara is geothermal situated in Ulumbu, Satarmese Sub-District, Manggarai District, precisely in the area of Volcanic Poco Leok, the south of Ruteng. The area comprises a 1000 Ha geothermal working area (*Wilayah Kerja Pertambangan,* WKP) which was under the management of PT PLN Persero since 1990.

This chapter examines the political economy of geothermal utilization in Geothermal Power Plant (PLTP) Ulumbu in Manggarai District, East Nusa Tenggara. The widening inequality in electricity in Indonesia is not related the sector itself, but also to other determinant factors, such as changes of institutions and conflicts of interests in the utilization of natural resources. Decentralization taking place in the late 1990s—through Law No. 22/1999, later replaced by Law No. 32/2004 on Regional Government—rought the development paradigm closer to the communities. It also sought to devise a compromise between conflicting interests in the utilization of natural resources (Agustina et al., 2012). At the end of 2014, the issuance of Law No. 23/2014 on Regional Government gave the authority back to the central government because of which energy and mineral resource affairs were once again delegated to the central and provincial authorities. Indirect utilization of geothermal energy for electricity generation was given also to the authority of the central government, as stipulated in Law No. 21/2014 on Geothermal which was under the same discussion as Law No. 23/2014.

This chapter is divided into four sections. The first section provides an overview of the disparities in electricity developments in Indonesia. The second describes the

M. T. Sambodo et al., *Electricity Access and Community Welfare in Indonesia*, SpringerBriefs in Energy, https://doi.org/10.1007/978-981-16-3611-0_3

utilization of geothermal energy in Indonesia. The third examines the conflicts of interests in the development of PLTP Ulumbu. The last summarizes and reflects on how this electricity inequality links with the political economy of the utilization of natural resources.

3.2 Portrait of Disparities in Electricity Developments

Manggarai District is part of East Nusa Tenggara Province, situated on each side of West and East Manggarai Districts and bordered with the Flores Sea to the north and the Savu Sea to the south. The district with a total population of around 327 thousand people and a per capita income around Rp. 2 million (USD 210) per month in 2012 engraves a portrait of inequality in electricity developments in Indonesia. As described in Chap. 1, there is a widening gap in the electricity developments in Indonesia, especially between Java and outside Java. Interestingly, low electrification ratios belong to regions with a great energy potential, such as Kalimantan, NTB, NTT, and Papua. Figure 3.1 shows very clearly how the paradox of electricity developments occurs in Indonesia, proving that proximity to energy sources does not guarantee a better electricity access for a given area. While Indonesia's national electrification ratio reached more than 76% in 2012, NTT only came at 46.5%. If we look it closely, the electrification ratio of Manggarai District was only around 34.5%. Satarmese

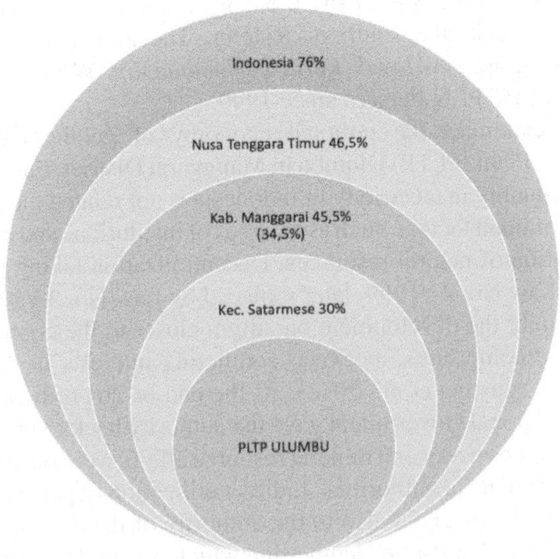

Fig. 3.1 Electrification ratio: portrait of development inequalities. *Note* Nusa Tenggara Timur (East Nusa Tenggara); Kab. Manggarai (Manggarai District); Kec. Satermese (Satarmese Sub-District). *Source* Sambodo et al. (2014)

Sub-district, which is home to the PLTP Ulumbu, had only an electrification ratio of around 30%. Even worse, the villages that are only a few kilometers from the PLTP had no access to electricity at all.

It is interesting to study more deeply the contradictory features of the electricity supply as shown in the figure above. It turned out that people who are living nearby the energy sources, like many others, were denied access to electricity. The political economy in the development of PLTP Ulumbu in Satarmese Sub-District, Manggarai District, may give a clear illustration of how this paradox is taking place. However, it is important to describe the context of geothermal developments in Indonesia before going into detail on the case of the PLTP Ulumbu.

3.3 Geothermal Utilization in Indonesia

Being a country with a large number of volcanoes, Indonesia is bestowed with a huge geothermal potential, even the largest in the world. There are at least more than 40% of the world reservation or 28.91 GW of geothermal potential, or similar to 11 billion Oil Barrel Equivalents (SBM), that can be developed in Indonesia for thirty years of operation (Geothermal Directorate, 2013; Water Resources and Mining Agency, 2011). However, only about 1.3 GW or 4.5% of the energy potential has been so far utilized. Seen from the energy mix used in Indonesia, geothermal energy contributed to just around 2.1% of the total energy consumption in 2010. The energy use has been now dominated by oil (48.4%), coal (24, 7%) and gas (19.7%) (Directorate of Geothermal Energy, 2013). The contribution of geothermal to the national energy mix is still far from the target as set in the National Energy Policy (KEN), which is 4% in 2015 and 7% in 2025.

The geothermal potential spreads evenly across Indonesia, from the western to the eastern part of the country. There are 90 geothermal sites in Sumatera, 71 in Java, 28 in Bali-Nusa Tenggara, 12 in Kalimantan, 65 in Sulawesi, 30 in Maluku, and 3 in Papua. Out of all these sites, as of September 2013, 58 geothermal mining working areas (WKP) have been established: 19 WKPs before Law No. 27/2003 has been into force and 39 WKPs after (Directorate of Geothermal Energy, 2013). Of the total 58 WKPs, only 9 WKPs have been operating, concentrated mostly in West Java, which became the first sites for geothermal development in Indonesia (see Table 3.1).

Historically, Kamojang geothermal site, about 42 km to the southeast of Bandung, West Java, marked the beginning of geothermal development in Indonesia. PLTP Kamojang was considered the first geothermal plant being developed for electricity generation. It stemmed from an exploration well found by the Dutch colonial government in 1926-1928. PLTP Kamojang Unit I began to operate in 1982 and was inaugurated by President Soeharto on February 7, 1983. Being the oldest geothermal development site in Indonesia, Kamojang becomes the main ground training for Indonesia's geothermal power (Department of Water and Resources, 2011).

There are at least two issues behind the underdeveloped geothermal power plants in Indonesia: financial and institutional. The first concerns the fact that developing

Table 3.1 PLTP installed capacity as of September 2013

No.	WKP location	PLTP	Developer/operator	Turbine capacity	Total capacity (MW)
1	Sibayak-Sina-bung, EAST SUMATRA	Sibayak	PT Pertamina Gethermal Energi	1 × 10 MW; 2 MW	12
2	Cibeureum-Parabakti, WEST JAVA	Salak	Chevron Geothermal Salak, Ltd	3 × 60 MW; 3 × 65.6 MW	377
3	Pangalengan, WEST JAVA	Wayang Windu	Star Energi Geothermal Wayang Windu	1 × 110 MW; 1 × 117 MW	227
4	Kamojang-Darajat, WEST JAVA	Kamojang	PT Pertamina Geothermal Energi	1 × 30 MW; 2 × 55 MW; 1 × 60 MW	200
5	Kamojang-Darajat, WEST JAVA	Darajat	Chevron Geothermal Salak, Ltd	1 × 55 MW; 1 × 94 MW; 1 × 121 MW	270
6	Dataran Tinggi Dieng, CENTRAL JAVA	Dieng	PT Geo Dipa Energi	1 × 60 MW	60
7	Lahendong-Tompaso, NORTH SULAWESI	Lahendong	PT Pertamina Geothermal Energi	4 × 20 MW	80
8	Waypanas, LAMPUNG	Ulubelu	PT Pertamina Geothermal Energi	2 × 55 MW	110
9	Ulumbu, NTT	Ulumbu	PT PLN	2 × 2.5 MW	5
Total					1341

Source Directorate of Geothermal Energy (2013)

geothermal energy is somewhat costlier than other power plants, such as Steam Power Plant (PLTU), Hydro Power Plant (PLTA), and Diesel Power Plant (PLTD). The prices for developing geothermal energy vary with an estimate of at least USD 2.8 million to USD 3.4 million per MW (Directorate of Geothermal Energy, 2013, 16). In addition to requiring high technologies, geothermal development is also much of a risk. The investigation of geothermal is not always successful and it is very possible to fail. Meanwhile, the exploration phase has a drilling success ratio at only 40–50%. (Directorate of Geothermal, 2013, 13–14). Because of the high risk, only big-money investors are capable of getting involved in the development of geothermal power plants for which they would obviously expect satisfactory electricity selling prices. For the present, the geothermal power prices still do not reach their economic values and, therefore, are less attractive to investors (Directorate of Geothermal, 2014).

The second issue relates to regulatory aspects of geothermal development. For instance, there are many overlaps in spatial planning between geothermal lands and forest areas. Before the enactment of Law No. 21/2014 on Geothermal, geothermal development should not be carried out in the protected and conservation forests. Consequently, this has created a serious impediment to the development of geothermal energy because around 15% (6000 MW) of the resources are situated in conservation areas and about 18% (7000 MW) in protected forests (Directorate of Geothermal Energy, 2014). The problem of overlap in land use has also complicated the licensing process, leading to the enactment of Law No. 21/2014. The newly commenced law stipulates that geothermal utilization is not considered a mining activity and is, therefore, allowed to take place in forest areas.

Currently, geothermal development in Indonesia has been under the regulation of Law No. 21/2014 on Geothermal. The House of Representatives (DPR) approved the law on August 26, 2014, replacing Law No. 27/2003 on Geothermal. If comparison should be made, it is evident that there is a huge difference between these two laws. First, the new Geothermal Law removed the term 'mining' for geothermal activities. That is, geothermal utilization is no longer considered a mining activity because it has different characteristics from mineral extraction. The activists of environment severely criticized the abolishment of the term for being 'a cheap trick' to allow geothermal utilization in forest areas.

Second, unlike the previous one, Law No. 21/2014 has a more detailed regulation on direct and indirect utilization. What is surprising is that under the new law indirect utilization of geothermal energy for electricity is now centralized. The provincial and district/city governments are only given the authority to deal with direct utilization of geothermal energy, such as for tourism. Whereas, the central government is fully controlling the indirect utilization of electricity generation, from determining the Geothermal Working Area (WKP) to issuing utilization permits. Indirect utilization of geothermal for electricity generation is considered strategic and supportive of national energy security that it should be concentrated under a single authority of the central government.

The informant from the Ministry of Energy and Mineral Resources said that the centralization of indirect utilization of geothermal was also because the human resources at the local governments are not yet ready to participate in the development of geothermal power plants (*Pembangkit Listrik Tenaga Panas Bumi,* PLTP). Moreover, the actual practice so far showed that the authority in the hands of the regional governments is considered an obstacle to the development of geothermal energy. It is believed that the central government has more skillful human resources to fully support the development of geothermal power plants.

It is interesting to note that centralization of indirect utilization of geothermal was absent in the Draft Bill (RUU) submitted beforehand by the government, but it somehow came out during the discussion session in the House of Representatives (*Dewan Perwakilan Rakyat,* DPR). On the one hand, the insertion of centralization into the draft could be interpreted as an effort to ensure the effectiveness and efficiency of accelerated development of geothermal power plants in Indonesia. On the other hand, it could be also considered against the spirit of decentralization and regional autonomy which have been mainstreamed in the reform era. Law No. 21/2014 on Geothermal does not stand on the spirit of regional autonomy that seeks to empower the local governments. The newly promulgated law also appears to contradict Law No. 30/2007 on Energy and Law No. 30/2009 on Electricity which push for more active roles of the local governments in providing energy and electricity.

The lack of support for the regional governments to participate in geothermal development is seen from how shares are distributed. The central government rejected the proposal that at least 10% of a participating interest should be mandatorily offered to the Regional-Owned Enterprise (BUMD). As a substitute, Law No. 21/2014 puts forward a production bonus to accommodate the regional interests. Article 53 states that geothermal permit holders are obliged to pay a production bonus to the local governments at a specified percentage of gross revenues since the first commercial operation of the unit. The regulation on a production bonus may be rather pleasing to the local governments, but at the same time it does not take as a priority the significance of their active involvements in geothermal development (Table 3.2).

Another important thing in Law No. 21/2014 on Geothermal is energy pricing. Article 22 stipulates that prices for electricity generated from the indirect utilization of geothermal by the government should take the economic values into account. This pricing policy guarantees profits to the investors involved in geothermal development and it, therefore, provides an incentive for further advancement. There is no doubt that institutional changes in Indonesia's geothermal management, as seen in the new Law on Geothermal, have a significant influence on the regional development of geothermal power plants. The following section will look at the dynamics and challenges of geothermal management in Satarmese Sub-District, Manggarai District, East Nusa Tenggara, and to what extent the institutional changes in geothermal management can address those challenges.

Table 3.2 Policy Implementation of geothermal activities

No.	Description	WKP before Law No. 27/2003	WKP after Law No. 27/2003	WKP after Law No. 21/2014
1	Operational Structures	JOC and ESC	Geothermal Mining Business Permit (IUP); Electricity Business Permit (IUK)	Geothermal Permit; Electricity Business Permit
2	Operational Patterns	*Total project* and *partial project*	*Total project*	*Total project*
3	Government's Share	34% of *Net Operating Income* (NOI) and tax	Tax and Non-Tax State Revenues (PNBP)	Tax PNBP, including a production bonus
4	Project Management	Pertamina (JOC) PLN (ESC)	IUP Holders	Geothermal Permit Holders

JOC joint operation contract; *ESC* energy sale contract
Source Elaborated from Directorate of Geothermal (2013)

3.4 PLTP Ulumbu and Regional Government's Hope

Ulumbu is the first Geothermal Working Area (WKP) that is directly controlled and operated by PLN. According to the PLN record (2014), the state-owned enterprise has started to manage the Ulumbu geothermal since 1990, based on the Letter of Pertamina Director of Operation and Production No. 2033/D0000/90-S1 dated December 27, 1990 on Small Scale Geothermal of WKP Ulumbu 1000 Ha. The letter was an answer to the PLN Letter No. 489/456/DIRUT/1990/M on Small Scale Geothermal of WKP Ulumbu, requesting for the WKP Ulumbu to Pertamina (PLN, 2014).

However, the development plan for the Ulumbu PLTP has been set out long before the PLN was assigned to control the geothermal site. The Mining and Energy Service of Manggarai District recorded that the Volcanology Directorate of Bandung initiated preliminary and follow-up investigations in 1975, 1978, 1984, and 1985. In 1989, KRTA Limited (New Zealand) did a Feasibility Study (FS) on Flores mini geothermal power project and Ulumbu geothermal field, while the University of Nusa Cendana carried out the Environmental Impact Analysis (AMDAL).

In cooperation with GEN-ZL and the New Zealand Ministry of Foreign Affairs and Trade, PLN drilled three exploration wells in 1994-1995 (PLN, 2014), including ULB-01, ULB-02, and ULB-03 located on the right ridge of the Wae Kokor River. In October 1996, GEN-ZL exerted the Feasibility Study Phase 2 on the Ulumbu mini geothermal power project (PLN, 2014).

According to PLN (2014), in the late 1990s the Asian Development Bank (ADB) began to get involved in the development of electricity in Manggarai. In 1999, the ADB fielded a fact finding mission to areas outside Java, including PLTP Ulumbu. In 2001, ADB Grant: TA No. 3481-INO for Outer Islands Electrification Project (including PLTP Ulumbu) was agreed. However, it was not until February 2008 that ADB signed a loan agreement for the construction of PLTP Ulumbu (2 × 2.5 MW). Following the ADB loan funds, a contract to build the PLTP Ulumbu (2 × 2.5 MW) through APBN funding program was signed in November 2011 (PLN,). Using an APBN-funded condensation technology with a capacity of 2 × 2.5 MW, PLTP Ulumbu began to operate in November 2011. Meanwhile, the ADB-funded power generation with a 2 × 2.5 MW back-pressure technology commenced its operation in mid-2014.

Despite its production since the end of 2011, The Government of Manggarai District considered that PLTP Ulumbu under the management of PLN has not fulfilled certain legal formalities. According to the provision of Law. No. 27/2003 on Geothermal, the PLN should have secured a Mining Business Permit (IUP) and Electricity Supply Business Permit as referred to in Law No. 30/2009 on Electricity. According to the Department of Energy and Mineral Resources, Manggarai District (2014), PLN has been running the PLTP Ulumbu based solely on a lawful Letter from the Pertamina Operation and Production Director in 1990 on Small-Scale Geothermal of WKP Ulumbu and the Letter from the Minister of Energy and Mineral Resources No. 3042/33/DJB/2009 on Small Scale Geothermal Activities of Ulumbu, NTT.

The fact that Manggarai District demanded the PLN to perform legal obligations is related to the wish for shared benefits from the electricity production of PLTP Ulumbu. Without the completion of legal formalities, it would be impossible to calculate the royalties or production fees for PLTP Ulumbu. It means also that the Government of Manggarai District, as the owner of Ulumbu geothermal potential, would lose potential incomes from this sector.

It seemed that the Government of Manggarai District has a sense of ownership towards the energy potential. In addition, the regional government claimed to have contributed to the construction of PLTP Ulumbu from 1991 to 2006, including land and plant acquisition amounting to Rp. 681 million, and of the 4.8 km Ponggeok-Ulumbu road equivalent of Rp. 9.3 billion, and building of Ruteng-Iteng road costing at Rp. 1 billion (Department of ESDM, Manggarai District, 2014).

In addition to the demand for an immediate realization of Non-Tax State Revenue (PNBP) from the electricity production of PLTP Ulumbu, the Government of Manggarai District also insisted that diesel fuel (BBM) savings, which have been previously used to run PLTD Ruteng, should be reallocated to expand rural electricity networks. The 77.6% or Rp. 65 billion reduction in fuel consumption in 2013 (Department of ESDM, Manggarai District, 2014), for example, could be used to expand additional networks as a means to improve the rural electrification ratios in Manggarai District.

To effectively deal with the problem of Ulumbu, Department of ESDM, Manggarai District (2014) made the following statement:

1. The PLN should pay much attention to the legal, economic, and socio-cultural aspects of Ulumbu geothermal management, both for the best interests of the PLN itself and the Government as well as the People of Manggarai District.
2. The Government of Manggarai District is supportive of the development programs for PLTP Ulumbu within a legal corridor.

3. The serious concern of Manggarai District with the management of PLTP Ulumbu is the manifestation of its regional responsibility and authority as mandated by Law No. 27/2003.

Furthermore, the Government of Manggarai District summoned the rights to participate in the management of the 5 MW PLTP Ulumbu. The President Director of PLN at the inauguration of the PLTP Ulumbu in 2011 made a promise to assure the participation of Regional-Owned Enterprises (BUMD), i.e. PT Manggarai Multi Investasi (MMI) (Department of ESDM, Manggarai District, 2014). The Government of Manggarai District schemed three scenarios for its participation. The first scenario is to welcome a pure private investor in the management of the PLTP. The private sector will serve as the operator developing and producing the electricity, then selling it to PLN. The participation of the Regional Government accounts for 5–10% of the investor's monthly income to be paid through the BUMD PT MMI. The second scenario is to set up a joint development project in which the BUMD and the investor are working together to develop PLTP Ulumbu, and sell the electricity produced to PLN. The third scenario is to let BUMD take a role, meaning that BUMD PT MMI will fully manage the still inactive 5 MW wells I and III (Department of ESDM, Manggarai District, 2014).

The Government of Manggarai District was desperate for participation in the management of Ulumbu geothermal. Therefore, it sent a letter to the Minister of Energy and Mineral Resources, requesting permission to allow PT MMI to conduct a Geothermal Preliminary Survey for the expansion of WKP Ulumbu. However, this effort was too late as the PLN made a faster move. On April 17, 2014, the Minister of Energy and Mineral Resources issued a Ministerial Decree No. 2363 K/30/MEM/2014 on Geothermal Preliminary Survey Assignment to PLN in Ulumbu Area, Manggarai District, NTT Province.

Law No. 21/2014 on Geothermal has been criticized for putting limits on the authority of the Government of Manggarai District. The Law was also thought to have discouraged BUMD from participating in geothermal energy development. The regional government's opportunity for taking a more active role in the management of geothermal power plants seems to have been severely shrinking due to the insufficient ability of BUMD PT MMI. Newly established in 2012, this BUMD is still lack of capital and human resources.

3.5 Conclusion

This chapter elucidates the Indonesia's geothermal utilization development in support of electricity improvement. In general, the utilization of geothermal energy for electricity in Indonesia is still relatively new. The oldest PLTP going into production for commercial purposes began to operate in the early 1980s. The enormous geothermal potential has not reached its optimal utilization in developing the electricity sector and reducing the disparities in electricity access in Indonesia. The development efforts still have to deal with a number of financial and institutional barriers.

The slow progress of the electricity sector has resulted in the birth of Law No. 21/2014 which is a revised form of the 2003 Geothermal Law. The newly enacted law stipulates that geothermal potential is of strategic importance for the national energy security that it should be under the authority of the central government. Pragmatically, the centralized authority of geothermal management for electricity development could be perceived as an effort to bring effectiveness and efficiency. However, this current law seems to have violated the spirit of regional autonomy in granting greater participation to the regional governments. There is also an impression that, unlike the Law on Energy and the Law on Electricity, the new law, in essence, wanted to undermine the roles of regional governments in improving regional electrification.

As for the case of PLTP Ulumbu in Manggarai District, there is obvious evidence that the regional government has a strong desire to get more involved in the development of geothermal potential in the area. The new Geothermal Law turned out to be irresponsive to the aspirations and it, therefore, has been considered against the best interests of the region. Amid alarming rates of electricity and poverty in Manggarai District, there is a need to consider the extent to which the regional government can participate in the geothermal energy development. Efforts to expand electricity networks to the rural areas, particularly the nearest villages to Ulumbu, are without a doubt the top priority.

The experience of community economic development around the areas of PLTP Kamojang and PLTP Darajat in West Java is a good lesson that tourism industry may become an alternative to increase the income of the neighborhoods. Tourism and geothermal generation industries can grow side by side provided that they are professionally managed. It is also obvious that a clear-cut development plan needs to be designed as early as possible to avoid conflicts of interests between the direct utilization of geothermal and the development of geothermal power plants (PLTP). Communication and coordination between the community, the local government, and the PLTP management are important keys for the two sectors to work together. In addition, the PLTP management needs to take advantage of the Corporate Social Responsibility (CSR) programs that can level up the people's income.

References

Agustina, C. D., Ahmad, E., Nugroho, D., & Siagian, H. (2012). Political economy of natural resource revenue sharing in Indonesia. Jakarta. Accessed on November 12, 2014 from http://www.lse.ac.uk/asiaResearchCen-tre/_files/ARCWP55-AgustinaAhmadNugrohoSiagian.pdf

Department of ESDM, Manggarai District. (2014). *Laporan tentang PLTPB Ulumbu, Kabupaten Manggarai, Nusa Tenggara Timur*. Manggarai.

Department of Water and Mining Resources. (2011). *Garut regency geothermal investment opportunities*. Kabupaten Garut.

Directorate of Geothermal. (2013). *Pengembangan panas bumi Indonesia*. Kementerian Energi dan Sumber Daya Mineral Republik Indonesia.

Directorate of Geothermal. (2014). *Pengembangan sektor panas bumi dalam upaya pengentasan pemiskinan.*

Minister of Energy and Mineral Resources of the Republic of Indonesia. (2014). Ministerial Decree of Energy and Mineral Resources No. 2363K/30/ MEM/2014 on Geothermal Preliminary Survey Assignment to PLN in Ulumbu Area, Manggarai District, NTT Province.

PLN. (2014). Historical Pengembangan Lapangan Panas Bumi Ulumbu (Presentation materials, unpublished).

Republic of Indonesia. (1999). Pemerintah daerah. *Undang-undang No. 22 tahun 1999 tentang Pemerintahan Daerah.*

Republic of Indonesia. (2003). Panas Bumi. *Undang-undang Republik Indonesia No. 27 tahun 2003 tentang Panas Bumi.*

Republic of Indonesia. (2004). Pemerintah daerah. *Undang-undang No. 32 tahun 2004 tentang Pemerintahan Daerah.*

Republic of Indonesia. (2007). Energi. *Undang-undang Republik Indonesia No. 30 tentang Energi.*

Republic of Indonesia. (2009). Ketenagalistrikan. *Undang-undang Republik Indonesia No. 30 tentang Ketenagalistrikan.*

Republic of Indonesia. (2014). Pemanfaatan panas bumi. *Undang-undang No. 21 tahun 2014 tentang Pemanfaatan Panas Bumi.*

Republic of Indonesia (2014). Pemerintahan daerah. *Undang-undang No. 23 tahun 2014 tentang Pemerintahan Daerah.*

Sambodo, M. T., Fuady, A. H., Negara, S. N., Handoyo, F., & Mychelisda, E. (2014). *Laporan Penelitian Model dan Strategi Pengembangan Sektor Ke- tenagalistrikan di Daerah dalam Upaya Pengentasan Kemiskinan.* Jakarta: Pusat Penelitian Kependudukan.

Chapter 4
Electricity Access and Community Welfare: A Case Study of Satarmese-NTT

4.1 Introduction

Improvement of access to electricity for the communities is one of the key agendas in the development of the electricity sector. In the five-year national development plan (2014–2019), the elected President, Joko Widodo, has set a target of electrification ratio at 100% in 2019 (Widodo & Kalla, 2014). It means that the government aims to make all the households in Indonesia have access to electricity in 2019. Following information from the Ministry of Energy and Mineral Resources, in 2019, the ratio of electrification was about 98.89% (ESDM, 2020). Many households do not have access to electricity because they live in remote areas. Improvement of access to electricity for communities is also one of the important pillars of poverty reduction programs. In the work plan of the National Team for the Acceleration of Poverty Reduction (TNP2K), advancement of community access to basic needs and services is identified as the top priority program one of which is through cheap and efficient electricity.

The expansion of electricity access to the communities as an effort to reduce poverty is in line with the sustainable development agenda. In the World Summit on Sustainable Development, in Johannesburg, South Africa in 2002, the world leaders spoke in one voice on the importance of sustainable energy development:

> to improve access to reliable and affordable energy services … (because) access to energy facilitates the eradication of poverty … this would include action to improve access to reliable, affordable, economically viable, socially acceptable, and environmentally sound energy services and resources … (UN, 2002, 11).

The world leaders agreed to work together to enhance energy access (including electricity) in support of poverty reduction. The UN declaration suggested further that improvement of access to energy should meet some criteria, i.e. reliable, affordable, economically viable, and environmentally sound. All these criteria are believed to guarantee the achievement of sustainable energy access.

Furthermore, Law No. 30/2007 on Energy, Article 3 (f) states:

M. T. Sambodo et al., *Electricity Access and Community Welfare in Indonesia*, SpringerBriefs in Energy, https://doi.org/10.1007/978-981-16-3611-0_4

... Improve accessibility to energy for the people who are less wealthy and/or who live in remote areas to bring about just and equal welfare and prosperity for the people...

That is, the poor people are also entitled to energy access. It is the task of the government (both central and regional) to make sure that their rights are fulfilled. The Energy Law came out earlier than the United Nations (UN) resolution No. 67/215/2012 aiming to bring energy access to all nations in the world. The UN resolution declared "The Decade of Sustainable Energy for All 2014–2024."

The rural areas suffer most the lack of access to electricity. International Energy Agency (IEA) (2014) showed that in 2012, the rural areas in Indonesia reached an average electrification ratio at only 59%, while the urban areas at 92%. The Indonesia's national electrification was at 76% in 2012. Within seven years, electrification ratio reached 98.89% in 2019, but, the electrification ratio in some provinces was still low, such as East Nusa Tenggara 85.85% (ESDM, 2020).

The Government, through the State-Owned Electricity Enterprise (PLN) continued with its plan to increase its rural electrification ratio, for which a special work unit has been established to handle the issue (PLN, 2014). The work unit (*Satker*) is assigned to run rural electricity programs, such as the APBN-funded installation of electricity networks (Laskari, 2012). Since 2011, the government has also budgeted Special Allocation Funds (DAK) for rural electricity development. In the previous years, the DAK was only referred to rural infrastructures, without a specific mention of allocation for the rural electricity sector. In 2011, the government allocated around Rp. 150 billion to support rural electricity programs. In 2012, this number was up to around Rp. 190.6 billion. Since 2013, the government has used the term 'DAK for rural energy' as a substitute for 'DAK for rural electricity'. This means that the scope of the program was broader and, as a result, the budget for rural energy got a significant rise at a total of Rp. 497.9 billion.

PLN has also sought to increase the number of electricity customers, especially in the rural areas. As shown in Fig. 4.1, it is roughly calculated that in 2022 there would be around 1.7 million new customers. Out of this number, the rural electricity customers are predicted to grow from around 220 thousand in 2013 to around 393 thousand customers in 2022. The government, through a cheap and economical electricity program, has subsidized around 95 thousand customers in 2013. The number presumably will decrease to around 89 thousand customers or an average of 3% of the PLN's new additional customers by 2022.

The growing number of the PLN customers resulted in the greater need for investment, both for the power plant construction and network distribution. According to the Electricity Supply Business Plan PT PLN, the average investment need for rural electrification is around 5% of the total PLN investment need (not including those under the scheme of private power generation or independent power producers) for 2013–2022. The investment need for rural electrification programs is going to go higher from around Rp. 2.9 trillion to Rp. 5.3 trillion in 2013–2022. This figure includes an investment cost of free electricity connections, as part of the rural electrification programs, which include investment cost of free electricity connection that is about Rp. 200 billion per year.

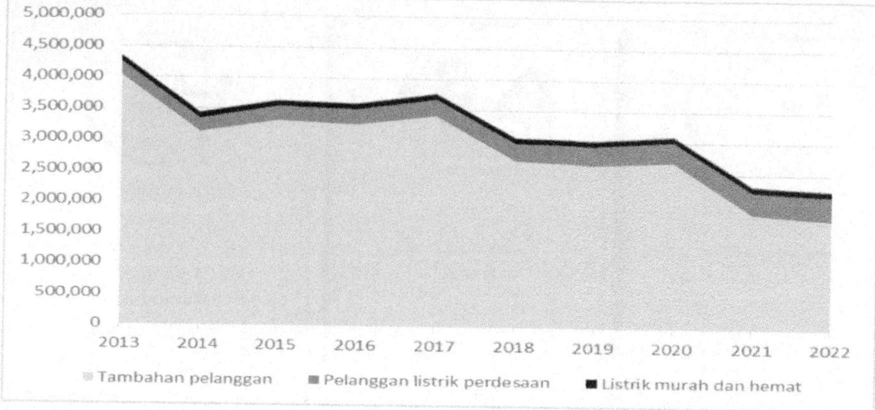

Fig. 4.1 Estimated number of PLN additional customers. *Note* Tambahan pelanggan (Additional Customers); Pelanggan listrik perdesaan (Customers of Rural Electrification); Listrik murah dan hemat (Cheap and Economical Electricity). *Source* PLN (2012)

As the Government vigorously attempts to spread out access to electricity, especially in the rural areas, an analysis is needed to measure the impacts of access to electricity on the people's welfare. This chapter seeks to show possible changes in the household welfare before and after access to electricity was gained. The people who were without access to electricity during the baseline survey but were no longer in the following year are called the treatment group. The people who were denied access to electricity are called the comparison or control group. Logically, the group with access to electricity would have a higher level of welfare than those without. By focusing its analysis on the changes of social welfare levels since access to electricity has been given, this study attempts to find out the extent to which access to electricity has impacts on the levels of people's expenditure, working hours, and study hours. Hopefully, this study can provide an empirical evidence on how provision of basic infrastructures, especially electricity, has benefited poverty reduction programs in rural areas.

4.2 A Literature Review on Electricity and Community Welfare

Very few of the researches in Indonesia studied quantitatively the impacts of access to electricity on the community welfare. One of the difficulties was inability to put aside the influencing factors other than access to electricity. Khandker et al. (2013), in particular, pointed out that it is extremely difficult to measure the direction and magnitude of causality between electrification programs and outcome indicators due

Fig. 4.2 Indicators of multidimensional poverty index (MPI). *Note* lama waktu sektolah (tahun)—Duration of School Period (Year); kehadiran di sekolah (School Attendance); kematian anak (Child Mortality); gizi (nutrition); listrik (electricity); sanitasi (sanitation); air (water); lantai (floor); bahan bakar untuk memasak (cooking fuels); kepemilikan aset (asset ownership). *Sources* Alkire et al., (2014)

to a complex linkage of various electrical devices, outputs, and intermediate outcomes (see Chap. 1).

Referring to the indicators of education, health, and quality of life as used in the Multidimensional Poverty Index (MPI), access to electricity is a supporting indicator for quality of life (see Fig. 4.2). Access to electricity in itself is a vital infrastructure to pursuit quality of education and public health.

A study by Alkire et al. (2014) found that the absence of access to electricity is one of the poverty causes (see Table 4.1). In many cases in Indonesia, access to electricity, along with other factors, contributes to poverty reduction. It is closely related to the efforts to achieve the Millennium Development Goals (MDGs). Electrical accessibility can also sustain productive economic activities. In addition, it can stimulate students' enthusiasm for learning; ease the workload from looking for firewood; reduce smoke pollution in the house; help dissemination of information through radio or television against various diseases; ensure environmental sustainability; and help the formation of global cooperation beneficial to development (UNDP, 2005).

Alkire et al. (2014) also suggested that poverty reduction efforts should be carried out in a comprehensive way. Table 4.1 shows that access to cooking fuels is a key factor for improving quality of life. Access to clean and cheap cooking fuels will reduce women's work burden in rural areas where they usually struggle to collect firewood for cooking fuels. Because they spend more time searching for firewood, they surely have less time for other productive activities. Besides, searching for firewood is full of risks and may endanger women's health.

Dakung et al. (1990) studied the impacts of electricity in Cisadane Village, Sukabumi. The research found that in addition to lighting, some households have

Table 4.1 Percentage of distribution of poverty causes

Country	1	2	3	4	5	6	7	8	9	10
	School Year	Child school attendance at	Rate of Child Mortality	Nutrition	Electricity	Sanitation	Drinking Water	Floor	Cooking fuels	Asset ownership
Thailand	29.2	11.5	19	12.2	1.2	4.8	4.4	2.5	10.6	4.6
Vietnam	18.5	14.3	12.9	12.2	1.5	12.1	5.5	5.5	13.1	4.4
Philippines	15.8	–	56.5	–	3.8	5.3	2.5	1.5	9.6	4.9
Indonesia	6.2	6.4	60.7	–	1.5	6.7	5.1	1.9	8	3.5
Lao PDR	16	15.4	18.9	11.5	6.3	9	5.3	2.3	10.9	4.4
Cambodia	14	8.1	13.5	19.3	10.9	10.6	6.8	1	11.9	4

Note school year = no one in the household completed a five-year school education; child school attendance = every school-age child were absent from school until grade 8; rate of child mortality = a number of children died in the family; sanitation = health facilities are not improved, or if it were, they are shared with other families; drinking water = the household is denied access to drinking water, or it takes more than 30 min (round trip) to find a safe drinking water; floor = the household has a ground- or sand-floored building; cooking fuels = the household are using firewood or charcoal for cooking; asset ownership = the household has no more than one radio, TV, telephone, bicycle, motorcycle or refrigerator, and has no motorcycle or car

Source: Alkire et al., (2014)

been using electricity to support productive business activities, including restaurants, beauty salons, dairy production, chicken farms, sewing, rice mills, lodging, and printing. It concluded that access to electricity has not only supported the already available productive economic activities, but also has opened up new business opportunities, such as sewing, ventures, restaurants, and beauty salons.

In regard with the benefits of electricity for the educational sector, Dakung et al. (1990) showed that access to electricity has boosted students' enthusiasm for learning. They spent more hours of learning at night and established groups of study. Accessible electricity has also increased participation in religious gatherings due to lighting and loudspeakers. Meanwhile, in the arts field, the presence of electricity as a source for lighting has supported art training and performances at night. Since gaining access to electricity, Cisadane Village has been also safer because participation in social or community activities, such as night patrols and other common activities in the village hall, is higher than before.

Before Dakung et al. (1990), BPS (1984) had carried out a survey on the impacts of electricity in the rural areas of North Sumatra and South Sulawesi Provinces. The survey revealed that there were at least five general patterns that can be inferred from the situation in North Sumatra and South Sulawesi Provinces. First, electricity use was limited to lighting, television and radio only because the power capacity accessible to most of the people was below 400 W. Second, in the educational sector, the school-age students spent more time for learning activities because of good lighting. Third, access to electricity multiplied the use of electronic devices, such as radio, television, and irons. Fourth, access to electricity escalated the productive business activities, particularly small sewing and businesses. Fifth, electricity used for lighting significantly decreased kerosene consumption for lighting fuels. Electricity dropped significantly kerosene consumption from an average of 16.2 L/month before to 3.9 L/month after. The use of electricity for cooking activities in the rural areas was still relatively limited. The rural communities still relied on firewood as a fuel for cooking, and very few of them used electrical equipment, such as rice cookers, water heaters, and electric stoves.

Access to electricity has a direct impact on the growing demand for both electronic goods, i.e. electric lights, radios, televisions, irons, fans, refrigerators, water heaters, rice cookers, air conditioners, and electric machines, i.e. sewing machines, lathes, and rice mills. The varied electronic devices are used to support different daily activities as well as to improve productivity and quality of work. Khandker et al. (2013) pointed out that the final product of electricity access is shown from the indicators of better education, income, and public health, the three of which interact with each other.

4.3 Descriptive Statistics and Econometric Models

Khandker et al. (2013) successfully identified the positive impacts of electricity access, particularly for women, in Vietnam's rural areas. According to the research, electricity access has saved the time spent to collect firewood and drinking water, prepare food, and grind rice, all kinds of activity being carried out by women in general. Using a panel data analysis, Khandker et al. (2013) measured the impacts of electricity access on levels of income, expenditure, school participation, and school completion. The results concluded that access to electricity has increased the household income and expenditure. The children also had more chances to participate in and graduate from their schools because of electricity. Accordingly, this condition will certainly help the children improve their welfare in the future for they will someday go into the labor market with a better level of knowledge.

This chapter is an attempt to apply the same econometric model as in Khandker et al. (2013). At first, this study tried to do an impact analysis of the World Bank-financed electrification program by drawing a comparison between the treatment and control groups. However, since the control group has been provided with electrification program funded by other sources during the study period, they no longer served as the comparison group for the treatment group. Alternatively, Khandker, Barnes, & Samad used a panel data estimation with a control variable of whether or not the households gained access to on-grid electricity for the purpose of which they applied a fixed effect analysis model.

The same problem came up also in this study wherein the selected control group turned out to have received access to electricity from the ministry's off-grid program. In addition, there was a comparison group who have been enjoying electrical power access for a much longer time. Three hamlets, i.e. Damu, Lungar/Mesir, and Tantong, were selected as the research sites where the impacts of electricity access on rural communities were studied (see Chaps. 1 and 2). The same households were surveyed in 2013 and 2014. When the baseline survey was taking place, Damu was the only hamlet having access to the PLN on-grid electricity in 2012. In 2013, there were only 20 households with access to metered electricity. In 2014, the number increased to 31 households (Table 4.2).

Some households still did not have access to the PLN metered electricity connections. According to them, an installation cost which is relatively higher than their income is the main reason for not having access to the PLN electricity. Some households could not afford to become PLN customers and, therefore, had to rely on the nearest neighbors with access to the PLN electricity.

In 2013, Tantong Hamlet had no access to electricity. Most of the households still relied on kerosene-fueled lamps for lighting. Geographically, Tantong Hamlet is adjacent to Damu Hamlet where a PLN electricity network is already available. Some people in Tantong Hamlet had to buy electricity from their fellows in Damu Hamlet. In 2013, 23 households in Tantong Hamlet gained access to unmetered electricity connections and 32 households had no access to electricity at all. In April 2014, PLN electricity networks entered Tantong Hamlet. A total of 48 households in Tantong Hamlet got connected to the PLN metered electrical grids. However, five households in Tantong Hamlet were still without access to the PLN electricity.

The pattern of access to electricity in Damu Hamlet and Tantong Hamlet was on-grid (with networks), while in Lungar/Mesir Hamlet was off-grid (see Table 4.2). The off-grid pattern in Lungar Hamlet was based on solar-powered lamps of the Super Extra Energy-Saving (SEHEN) with a capacity of around 12 Wp (Watt peak).

Table 4.2 Condition of electricity access in three villages/Hamlets (Number of household)

Village/Hamlet	2013	2014
Damu-on-grid		
PLN-metered	*20*	31
PLN-unmetered	*23*	11
Without access to electricity	*0*	1
Total Household	*43*	43
Tantong-on-grid		
PLN-metered	*0*	45
PLN-unmetered	*24*	3
Without access to electricity	*28*	4
Total Household	*52*	52
Lungar-off-grid-SEHEN		
Connected	*82*	105
Unconnected	*105*	69
Disconnected	–	13
Total Household	*187*	187

Note In 2013, there was one family in Damu Hamlet who had access to unmetered electricity connection. However, in 2014 this family had no longer access to both metered and unmetered electricity connections
Source Author (Primary data)

In 2013, 82 out of the 187 households were recorded to have access to electricity through SEHEN. In 2014, the SEHEN customers grew to 105 households. However, 13 households have lost their connections in 2014. There were two reasons behind this cutoff: either they could not afford to pay monthly contributions so that the PLN had to disconnect the installed SEHEN devices, or they voluntarily left the SEHEN program because of its lower-than-expected quality.

As mentioned in the previous section, access to electricity can reduce energy expenditure, especially on kerosene. This was also the case in Tantong Hamlet wherein the average rate of kerosene expenditure has dropped from Rp. 38,270/month to Rp. 23,173/month or a decrease at around 39% after the availability of electricity access (see Table 4.3). At the same time, the electricity connection costs have resulted in a significant rise of more than 100% in the household electricity expenditure, especially in Damu Hamlet. Every new customer was obliged to bear a fee between Rp. 2 to 3 million contingent upon the installed power capacity. The connection fee was fixed (i.e. paid once only) after which the electricity consumption cost is mainly determined by the amount of power used.

It is interesting to note that expenditure on fuel wood was on a significant increase in Tantong and Damu. The total energy expenditure in Tantong Hamlet was up by 24.5%. At the same time, the hamlets without access to electricity, i.e. Lungar/Mesir, showed a decline in kerosene expenditure by 6% probably because of the additional

Table 4.3 Average food and non-food expenditures/month in Rupiah (Rp.)

Type of expenditure	Tantong		Damu		Lungar/Mesir	
	2013	2014	2013	2014	2013	2014
Foods	185,923	136,750	136,012	181,000	196,508	297,123
Kerosene	38,269	23,173	25,116	31,628	27,770	26,117
Electricity	11,346	23,574	39,442	40,093	24,521	25,513
Fire wood	87,115	123,462	38,198	152,093	74,225	45,109
Non-foods and energy	221,198	324,444	277,884	550,016	354,383	330,656
Total household expenditure	543,852	631,403	516,651	954,829	677,407	722,678
Total per capita expenditure	102,614	140,312	103,330	190,966	120,966	129,050

SEHEN customers. In contrast to Tantong Hamlet and Damu Hamlet, Lungar/Mesir Hamlets showed a downward trend in firewood expenditure. A considerable increase in firewood expenditure occurred when some families were performing ritual meals or other festivals that require a lot of firewood for cooking. Thus, it is save to conclude that firewood expenditure remained at the top rank of energy expenditure.

There was a high hope that access to electricity would escalate the level of welfare measured from the rate of household expenditure. The household expenditure is representative of income. Thus, an increase in the household spend was expected to eventually move more households out of the poverty line. If we refer to the poverty line for rural areas as set by East Nusa Tenggara Province (i.e. Rp. 217,918/capita/month in March 2013; and Rp. 248,606 in March 2014), we can conclude that the average per capita expenditure in the three hamlets remained below the poverty line. Even though there has been an increase in per capita expenditure between 2013 and 2014, it was still insufficient to lift the people's average expenditure over the poverty line.

The next question is whether the on-grid or off-grid electricity access (SEHEN) has contributed to the higher rate of household expenditure above (SEHEN). Our preliminary observations showed that due to the available access to electricity most of the people have been able to save much from their kerosene expenditure for lighting. Table 4.3 shows the average kerosene expenditure in Tantong Hamlet reached a lower rate at Rp. 15,000/month. Considering the kerosene price at Rp. 4,000/liter, the average kerosene consumption for each household has dropped by 3.75 L/month/family. Some residents in Tantong Hamlet informed that the saved money from the decline in kerosene expenditure was spent in exchange for foods and non-foods.

This study used two approaches when measuring the impacts of electricity access on the community welfare. The first was a panel data analysis as in Khandker et al. (2013), aimed at grasping the impacts of on-grid and off-grid electricity access on the public welfare. The second was a Difference in Difference (DID) method of measurement for the impacts of electricity access in Tantong Hamlet in particular. Because some households have obtained on-grid and off-grid electricity access, the households with no electricity in 2013 and 2014 were selected as the control group.

4.4 Econometric Model: Panel Data

This study used a model of panel data analysis as in Khandker et al. (2013) with the following formulation:

$$Y_{ijt} = \beta^y x_{ijt} + \gamma^y V_{jt} + \delta_h^y E_{Hijt} + \delta_k^y E_{Kijt} + \delta_v^y E_{Vjt} + \chi^y T_t + \varepsilon_{ijt}^y \quad (4.1)$$

where Y_{ijt} is the output (total real expenditure deflated by the consumer price index)[1] of household *i* in hamlet *j* in year *t*; E_{Hijt} is on-grid electricity access (1 if household *i* in hamlet *j* has electricity connection and 0 otherwise)[2]; E_{kijt} is off-grid electricity access/SEHEN (1 if household *i* in hamlet *j* has electricity connection and 0 otherwise); E_{vjt} is a status of on-grid electricity access at the hamlet level (1 if hamlet *j* has on-grid electricity network connection and 0 otherwise); X_{ijt} is the characteristic at the household level, i.e. number of family members, access to *Raskin* (the rice for the poor), and floor area; V_{jt} is the hamlet characteristics (the presence of PLN electricity network); and *t* represents the time period (2013 and 2014); β^y ,γ^y , δ_h^Y , δ_k^Y and δ_v^Y are the parameter estimate from Eq. (4.1); ε_{ijt}^y and is a non-systematic error.

To have a direct estimate of Eq. (4.1) would be problematic because the E_{Hijt}, H_{Kijt}, E_{Vjt} and Y_{ijt} are simultaneously determined by a group of characteristics, both observed and unobserved. For example, the household's decision to have electricity connection is not only influenced by the infrastructure condition in the village, especially the PLN electricity network (V_{jt}) but also by the household's characteristics (X_{ijt}). During the baseline survey, we often found that the household could not afford to pay the connection fee because it is fairly expensive and beyond their income. This condition was not only because of the low income level, but also the large number of family dependents. Thus, the equation for the household electricity connection (on-grid and off-grid) can be formulated as follows:

$$E_{Hijt} = \beta^e x_{ijt} + \gamma^e V_{jt} + \chi^e T_t + \varepsilon_{ijt}^e \quad (4.2)$$

[1]CPI (Consumer Price Index) refers to the condition of the NTT province between June 2013 and 2014, the inflation rate was 8.31%; CPI in June 2014 was 113.63 (the constant prices in 2012) and CPI in 2013 was 104.9.

[2]It is important to note that the household's access to electricity is either sourced from the PLN (on-grid) or through the PLN SEHEN program (off-grid). Both have very different characteristics, particularly in term of power production, which surely affect the level of electricity utilization. Therefore, it is of significance to measure separately the effects of SEHEN electricity access.

$$E_{Kijt} = \Phi^f x_{ijt} + \Phi^f V_{jt} + \chi^f T_t + \varepsilon^f_{ijt} \tag{4.3}$$

Likewise, the equation at the hamlet level can be written as follows:

$$E_{vjt} = \gamma^v V_{jt} + \chi^v T_t + \varepsilon^v_{ijt} \tag{4.4}$$

Thus, for the outcome equation, ε^v_{ijt} is a combination of three error terms component:

$$\varepsilon^y_{ijt} = \mu^y_j + \eta^y_{ij} + e^y_{ijt} \tag{4.5}$$

where $\mu_j{}^y$ and $\eta_{ij}{}^y$ represent the unobserved hamlet condition and the unobserved household characteristics, while e^y_{ijt} is a non-systematic error that has no correlation with the other two errors. Meanwhile, the error components in Eqs. (4.2), (4.3), and (4.4) can be presented as follows:

$$\varepsilon^e_{ijt} = \mu^e_j + \eta^e_{ij} + e^e_{ijt} \tag{4.6}$$

$$\varepsilon^f_{ijt} = \mu^f_j + \eta^f_{ij} + e^f_{ijt} \tag{4.7}$$

$$\varepsilon^v_{jt} = \mu^y_j + e^y_{jt} \tag{4.8}$$

Since there is a possible correlation between ε^y_{ijt}, ε^e_{ijt}, ε^f_{ijt}, and e_{jt}, the variables $E_{Hijt}, E_{Kijt}, E_{vjt}$ and Y_{ijt} will likely to connect with each other due to the unobserved factors at the hamlet level and the household characteristics, causing a problem of endogeneity. For example, the hamlets with good road access are found to have more on-grid electricity access and, therefore, are given the highest priority for electricity access compared to those in the remote areas. Similarly, when a hamlet is given access to electricity, more households have more open opportunities for economy as well as the ability to have electricity connection. These two are sources of an endogeneity problem that will possibly cause a bias toward the parameter estimate of the impacts of electricity access on the output variables.

The problem of endogeneity can be avoided through a panel data analysis using a fixed-effect method. This model assumes that unobserved heterogeneity tends to remain the same throughout a period of observation at the household and village levels. For a very short period of estimation, this assumption is quite reasonable. Using a Fixed-Effect Model that can eliminate the effect of unobserved heterogeneity, Eq. 4.1 can then be written as follows:

$$\begin{aligned} Y_{ij1} - Y_{ij0} &= \beta^y (X_{ij1} - X_{ij0}) + \gamma^y (V_{j1} - V_{j0}) + \delta^y_h (E_{Hij1} - E_{Hij0}) \\ &\quad + \delta^y_k (E_{Kij1} - E_{Kij0}) + \delta^y_v (E_{Vj1} - E_{Vj0}) + \chi^y (T_1 - T_0) + \left(\varepsilon^y_{ij1} - \varepsilon^y_{ij0}\right) \end{aligned}$$

or

$$\Delta Y_{ij} = \beta^y \Delta X_{ij} + \gamma^y \Delta V_j + \delta_h^y \Delta E_{Hij} + \delta_k^y \Delta E_{Kij} + \chi^y \Delta T + \Delta\varepsilon_{ij}^y \tag{4.9}$$

Equation (4.9) gives an unbiased estimate provided that the time-invariant heterogeneity is fulfilled. However, the time-invariant assumption of heterogeneity may fail for several reasons. First, the unobserved factors that affect the outcome variables of household and hamlet with and without electricity may have changed. For example, the timing for on-grid and off-grid electricity network connections may vary across the households and hamlets. Second, relatively remote hamlets may have some delay in access to electricity compared to those with better road access. Third, some households may take the advantage of connecting to electricity earlier for they have adequate financial capabilities or they discern the large potential profits when connected to the electricity network. Thus, differences in the timing of access to electricity connections, characteristics of respondents, and unequal hamlets, have an effect on the dynamics of electricity connections and different growth projections.

Under the condition of time-variant heterogeneity, the error structure on Eq. (4.5) can be written as follows:

$$\varepsilon_{ijt}^y = \mu_{jt}^y + \eta_{ijt}^y + e_{ijt}^y \tag{4.10}$$

So, Eq. (4.9) can be written as follows:

$$\begin{aligned}\Delta Y_{ij} = \beta^y \Delta X_{ij} + \gamma^y \Delta V_j + \delta_h^y \Delta E_{Hij} + \delta_k^y \Delta E_{Kij} \\ + \chi^y \Delta T + \Delta\mu_{ij}^y + \Delta\eta_{ij} + \Delta e_{ij}^y\end{aligned} \tag{4.11}$$

where $\Delta\varepsilon_{ij}^y = \Delta\mu_{ij}^y + \Delta\eta_{ij}^y + \Delta e_{ij}^y$ will correlate with access to electricity. Under this condition, the ordinary least square (OLS) on Eq. (4.11) will produce inconsistent estimates. One way to solve the problem is to take into account the relationship between unobserved heterogeneity, initial conditions of the household, and characteristics of the hamlet. The initial characteristics of the hamlet will determine its prospects for access to electricity. These characteristics will also give different responses at the household level. Thus, Eq. (4.9) can be rewritten as follows:

$$\begin{aligned}\Delta Y_{ij} = \beta^y \Delta X_{ij} + \gamma^y \Delta V_j + \delta_h^y \Delta E_{Hij} + \delta_k^y \Delta E_{Kij} \\ + \delta_v^y \Delta E_{vj} + \alpha_h^y \chi_{ij0} + \alpha_v^y V_{jo} + \chi^y \Delta T + \Delta e_{ij}^y\end{aligned} \tag{4.12}$$

Equation (4.12) may result in an unbiased estimate.

4.5 Estimation Results of Panel Data

The results of panel data analysis for Eq. 4.12 are shown in Table 4.4 from which we can draw at least three main conclusions. First, the availability of electricity access through the PLN (on-grid) network statistically has no effect on the output variables,

Table 4.4 Estimation results of panel data

Dependent Variables				
Independent variables	Change in total expenditure	Change in food expenditure	Change in kerosene expenditure	Change in non-food and energy expenditure
	Coefficient	Coefficient	Coefficient	Coefficient
On-grid	−0.039 (0.225)	−0.132 (0.269)	−0.136 (0.224)	−0.128 (0.316)
Off-grid	−0.458 (0.194**)	0.006 (0.241)	−0.510 (0.409)	−0.683 (0.273**)
Rural electricity networks	0.446 (0.252***)	−0.039 (0.303)	−0.451 (0.263***)	0.852 (0.357**)
Rice for the poor (Raskin)	0.0259 (0.115)	0.218 (0.141)	−0.194 (0.144)	0.164 (0.163)
Number of family member	0.111 (0.083)	0.190 (0.1005***)	−0.090 (0.098)	0.142 0.118
Floor space	−0.146 (0.193)	0.008 (0.232)	−0.034 (0.247)	−0.378 0.273
Initial condition of number of household	−0.034 (0.023)	−0.074 (0.027***)	0.022 (0.030)	0.0014 0.032
Initial condition of Raskin	−0.249 (0.207)	−0.143 (0.248)	−0.033 (0.239)	−0.286 0.292
Initial condition of on-grid	0.207 (0.242)	0.187 (0.289)	0.181 (0.256*)	0.181 0.34**
Initial condition of off-grid	0.167 (0.115)	−0.108 (0.141)	0.162 (0.216)	0.402 0.162***

(continued)

Table 4.4 (continued)

Dependent Variables				
Initial Condition of rural electricity network	−0.765 (0.251*)	−0.254 (0.301)	−0.433 (0.263)	−0.609 0.354
Initial condition of floor space	0.0007 (0.003)	0.003 (0.004)	0.005 (0.004)	−0.001 0.005
Time	−0.378 (0.255)	−0.361 (0.311)	0.300 (0.288)	−0.183 0.362
Number of observation	533	527	381	532
Intra-class variation-rho	0.529	0.395	0.543	0.595

s.e. standard error
*significant at 1%; ** significant at 5%; ***significant at 10%

i.e. changes in the total expenditures for foods, kerosene, non-foods and energy. This is a quite contrast to the initial hypothesis that access to electricity will affect the household spending patterns Khandker et al. (2013). This result emerged because the electricity access received by the people of Tantong Hamlet started only within less than three months (from the time when electricity was installed until the survey was conducted). As a consequence, the impacts of electricity access have not been yet significant. The same case applied also to Damu Hamlet wherein some residents only have access to metered electricity.

Second, the off-grid electricity access, i.e. SEHEN, has proved to decrease the total expenditures for non-foods and energy. This indicates that having access on SEHEN reduce the total expenditure on the average about 58%.[3] As stated in the previous section, the SEHEN electricity access has been only enjoyed by some residents in Lungar/Mesir Hamlets. The PLN has set some criteria to determine who first should be given access to the SEHEN. One of the criteria said that the priority will be given to those residents who have a purchasing capability and are located far from the PLN networks. The Lungar/Mesir Hamlets have been beyond the PLN networks and there is still no plan up to now to build an electricity network in the two hamlets. Given that access to the SEHEN has reduced the growth of the total expenditure as well as spending for non-foods and energy, it is true to claim that the program has just benefited some households. It is interesting to see that the magnitude of change is higher than the total change in expenditure. This implies that, household with SEHEN have better energy efficiency and they can reduce the energy cost. When the survey was conducted in the first and second years in Lungar/Mesir Hamlets, some residents expressed their high hopes for PLN electricity access through the SEHEN program.

Third, when the electricity network has been entering the hamlets, there was an increase in total expenditure growth as well as spending for non-foods and energy. This indicates that the rural electricity network increase total expenditure by 56.2% on average than rural without electricity network. Damu Hamlet and Tantong Hamlet are the only areas that have access to the PLN electricity network and they have far better road infrastructures than Lungar/Mesir Hamlets. In this context, an increase in expenditure is not solely determined by the presence of the electricity grids but also the condition of other infrastructures.

Since Damu Hamlet and Tantong Hamlet received access to electricity (rural electricity network), the two areas had a lower average expenditure on kerosene at about 57% (ceteris paribus). In this sense, access to the PLN electricity has a positive correlation with a decrease subsidy in kerosene. Further, rural electricity network has increased non-food and energy expenditure on average by 134.4% (ceteris paribus). We argued that this is due to increasing spending on electric installation and electronic equipment such as radio, television, fan and many others.

---

[3]Because we expressed independent variable in log and dependent variable in dummy, we need to transform the parameter as follows: 100 x ($e^{\text{paremeter estimate}}$ -1).

4.6 Model of Difference in Differences

This section seeks to measure the impacts of PLN on-grid electricity access in Tantong Hamlet as a treatment group in comparison with Lungar/Mesir Hamlets as a control group. Lungar/Mesir Hamlet was chosen as the control group because some residents in the hamlets have not yet gained on-grid and off-grid electricity access. As for the treatment group in Tantong Hamlet, the team selected households that have access to electricity only in 2014. The households with unmetered electricity access through their neighbors were excluded from this analysis.

The approach of Difference in Differences (DID) combines two aspects in choosing the control groups (counterfeit of counterfactuals), i.e. the comparison between before and after the treatment, as well as between participants and non-participants in the program (Gertler et al., 2011). Compared to pre- and post-program approaches, the DID model provides a better estimate. Nevertheless, in order for the DID model to obtain a higher level of accurate estimation, the control group must accurately represent the same change condition as the treatment group when there is no project (Gertler et al., 2011).

As shown in Fig. 4.3, those poor families who are either with or without access to electricity underwent an increase in expenditure (as represented by the two parallel lines). To measure more accurately the impacts of the electrification program, it is necessary to take into account the differences between the two households in each time period. By doing so, the net changes between the two time periods can be successfully observed. In the econometric model, the DID analysis is formulated as follows.

$$H_{it} = \alpha + \beta.1(t = 1) + \gamma.1(D_1 = 1) + \delta.1(t = 1).1(D_i = 1) + \varepsilon_{it} \quad (4.13)$$

where $i = 1, \ldots,$ N households, and in expectation of two time periods, Eq. (4.13) can be written as follows:

$$E(H_{i1}|D_i = 1) = \alpha + \beta + \gamma + \delta$$

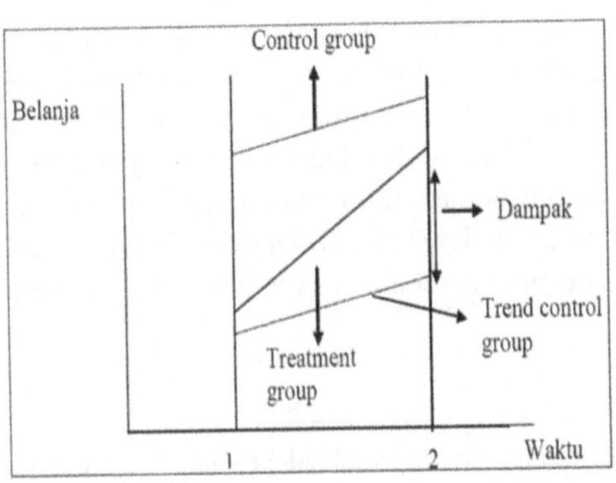

Fig. 4.3 Measurement of changes in welfare Levels. *Note* Belanja (expenditure); dampak (impact); waktu (time). *Source* Sambodo et al. (2014)

$$E(H_{i0}|D_i = 1) = \alpha + \gamma$$
$$E(H_{i1}|D_i = 0) = \alpha + \beta$$
$$E(H_{i0}|D_i = 0) = \alpha \qquad (4.14)$$

The DID parameter, which evaluates the impacts of electrification on the indicator of the community welfare level, is obtained by the following formula:

$$DD = [E(H_{i1} : D = 1) - E(H_{i0} : D = 1)] - [E(H_{i1} : D = 0) - E(H_{i1} : D = 0)]$$
$$DD = \{[\alpha + \beta + \gamma + \delta] - [\alpha + \beta]\} - \{[\alpha + \beta] - \alpha\}$$
$$DD = \delta \qquad (4.15)$$

where δ is the parameter that measures changes in the expenditure level; $E(H_{i1}$: $D = 1)$ is the parameter value of expenditure of the households participating in the electricity connection program in the second year; $E(H_{i0}$: $D = 1)$ is the parameter value of expenditure of the households participating in the electricity connection program in the first year; $E(H_{i1}$: $D = 0)$ is the parameter value of expenditure of the households participating in the electricity connection program in the second year; and E $E(H_{i0}$: $D = 0)$ is the parameter value of expenditure of the households not participating in the electricity connection program in the first year.

After that, the desired parameters are to be obtained through the following econometric model estimation:

$$outcome_{var_i} = \beta_0 + \beta_1.period_i + \beta_2.treated_1 + \beta_3.period_i.treated_i + \beta_k x_{k,i} + \varepsilon_i \qquad (4.16)$$

where each parameter is interpreted as follows:

$\hat{\beta}_0$ is the average output of the control group in the baseline period.

$\hat{\beta}_0 + \hat{\beta}_1$ is the average output of the control group in the period after having access to electricity.

$\hat{\beta}_2$ is the difference between the treatment and control groups in the baseline period.

$\hat{\beta}_1 + \hat{\beta}_2$ is the output average of the treatment group in the baseline survey.

$\hat{\beta}_0 + \hat{\beta}_1 + \hat{\beta}_2 + \hat{\beta}_3$ is the average output of the treatment group after having access to electricity

$\hat{\beta}_3$ is DID estimator representing the impacts of the availability of electrical connection.

Table 4.5 shows the estimation results of the DID analysis by taking into account four output indicators, i.e. total expenditure, food expenditure, kerosene expenditure, and non-food and energy expenditure. Access to electricity has increased the total household expenditure by 74.5% (ceteris paribus) in Tantong Hamlet.[4] That is, there was an increase in the total expenditure in the treatment group after having access

[4]Because we expressed independent variable in log and dependent variable in dummy, we need to transform the parameter as follows: 100 x ($e^{\text{paremeter estimate}}$ -1).

Table 4.5 Estimation results of DID model

Independent variables	Dependent variables			
	Total expenditure	Food expenditure	Kerosene expenditure	Non-food and energy expenditure
	Coefficient	Coefficient	Coefficient	Coefficient
Period	−0.052 (0.097)	0.06 (0.120***)	−0.034 (0.076)	0.087 (0.14)
Treatment	−0.456 (0.154*)	−0.322 (0.163)	−0.259 (0.103**)	−0.544 (0.23**)
Impact	0.557 (0.188*)	−0.104 (0.215)	−0.412 (0.196**)	0.749 (0.28*)
Floor space	0.169 (0.118)	0.347 (0.153**)	0.083 (0.106)	0.113 (0.17)
Rice for the poor (Raskin)	−0.164 (0.125)	0.233 (0.157)	−0.278 (0.127**)	0.029 (0.2)
Family Member	0.122 (0.02*)	0.114 (0.021*)	0.058 (0.016*)	0.153 (0.03*)
Number of observation	219	218	196	218
Adjusted R_2	0.2279	0.2085	0.2484	0.1674

s.e. standard error
*significant at 1%; ** significant at 5%; ***significant at 10%

to electricity. However, access to electricity did not affect food expenditure. The DID estimation results also showed that access to electricity has reduced kerosene expenditure about 51% (ceteris paribus) in the treatment group. Lastly, access to electricity has also aggregated non-food and energy expenditure by 75% (ceteris paribus).

The analysis results of the DID method brought a more rigorous conclusion regarding the impacts of access to electricity on household expenditure. The reason is that DID model has only assessed the households in Tantong hamlet with access to electricity in 2014 (or the treatment group) and those in Lungar/Mesir hamlets with no access to (both on-grid and off-grid) electricity. The panel data and the DID approaches showed a decline in the household kerosene consumption after electricity access has been given. This implies that access to electricity may also support the efforts to reduce energy subsidies.

4.7 Direct Impacts of Electricity Access

Table 4.6 shows that the households without access to electricity tended to spend more money on buying fuel wood and kerosene, and vice versa for those with access to electricity. This also indicates that access to electricity can sustain fuel subsidy reduction programs, especially on kerosene.

Table 4.7 also makes a comparison of different levels of welfare between the households with and those without access to electricity. In this case, the research team tried to examine the differences in the levels of food and non-food expenditure, total expenditure, total working hours and total study hours between the two groups during the 2013–2014 period. In term of expenditure, all the households (both with and without access to electricity) experienced an increase in food and non-food expenditure. Interestingly, non-food expenditure among the households with access to electricity has tripled during the observation period.

We learn from Table 4.7 that it is rather difficult to measure the community welfare based on the indicators of food and non-food expenditures. This is because there are many factors other than electricity access that can affect household expenditure. Moreover, people in rural areas typically belong to close kinship ties or mutual cooperation because of which they have a tendency to help each other in covering their daily needs. Therefore, the research team turned to other indicators, such as hours of work and study, to gauge the impacts of electricity access on the households.

The indicators of total hours of work and study showed that the households with access to electricity spend more and longer hours during the 2013–2014 period. On the contrary, the households without access to electricity suffered a decline in the two indicators. It is disconcerting that during the observation period there was a significant decrease in study hours among the children of the households without access to electricity. An increase in hours of work and study is one of the important findings in this study in regard with the impacts of access to electricity on the rural communities.

Table 4.6 Average expenditure of fuelwood and kerosene of households with and without access to electricity, 2013–2014 (Rp per month)

Indicator	2013		2014	
	With electricity access	Without electricity access	With electricity access	Without electricity access
Expenditure for fire food (Rp)	9.637	108.684	144.222	160.625
Expenditure for kerosene (Rp)	30.602	45.752	31.967	47.304
Number of household	188	119	237	107

Note data from the amalgamation of all hamlets in Satarmese, including the hamlets of Tantong, Lungar and Damu

Source: primary data

Table 4.7 Comparison of the welfare level of households that have and do not have access to electricity (at mean value level), 2013–2014

Indicator	2013		2014	
	With electricity access	Without electricity access	With electricity access	Without electricity access
Total food expenditure (Rp /month)	185.051	204.748	251.240	226.471
Total non-food expenditure (Rp/month)	587.796	428.081	716.328	1.257.043
Total expenditure (Rp/month)	772.846	636.950	961.776	1.493.284
Total hour of working (in a month)	154,1	144,5	154,7	141,9
Total hour of studying (in a day)	4,6	4,5	4,7	2,9

Note data from the amalgamation of all hamlets in Satarmese, including the hamlets of Tantong, Lungar and Damu
Source: primary data

4.8 Conclusion

This chapter attempts to measure the impacts of access to electricity on people's welfare based on various indicators, i.e. expenditures on food, on non-foods, on kerosene and fire wood. Using a panel data analysis, it found that the average communities in the three hamlets studied are still living below the provincial poverty line. Access to electricity in the three hamlets varies in characteristics, from on-grid to off-grid, and metered to unmetered.

The fixed-effects approach of panel data overcame the problem of unobserved heterogeneity. The model was able to figure out the impacts of on-grid and off-grid electricity. The DID approach focused solely on the measurement of the impacts of on-grid electricity access on the treatment group. Both approaches were used complementarily.

The panel data estimates demonstrated that on-grid electricity access has no impact on the household expenditure. This results is different with Khandker et al. (2013) who argued that on-grid electricity access has positive impact on household income and education. We argue that this is due to very short impact assessment. Meanwhile, off-grid electricity access (SEHEN), that have been developed longer than on grid, led to a decline in the total household expenditure and consumption on non-foods and energy. Similar with Khandker et al. (2013), we argue that even off grid connection can bring positive impact on households. Thus, access to the SEHEN in Lungar/Mesir hamlets needs to continue and reach the households who have not become the customers. It is still uncertain when the two hamlets will gain access to the PLN grid.

In the meantime, the presence of rural electrification network has a positive impact on the significant growth of total expenditure and spending on non-foods and energy. Yet, this result should be interpreted cautiously because the impacts of the construction of other infrastructures before access to electricity, such as road access, might affect the estimation results. We found, nevertheless, a 45% (ceteris paribus) reduction of the average kerosene expenditure after the electricity network has been available in Damu Hamlet and Tantong Hamlet.

The results of DID analysis showed that the treatment group, i.e. the households with access to electricity in Tantong Hamlet, has experienced a significant increase in non-food and energy expenditure and, at the same time, a considerable decrease in kerosene spending. Despite of a very short evaluation period (over less than four months after the communities gained access to electricity), the positive impact of access to electricity on improving the community welfare, based on the measured changes in the household expenditure, has demonstrated the desired outcome.

References

Alkire, S., Conconi, A., & Seth, S. (2014). *Multidimensional poverty index 2014: Brief*. Accessed on August 20, 2014 from http://www.ophi.org.uk/wp-content/uploads/Global-MPI-2014-Brief-Methodological-Note-and-Results.pdf

BPS. (1984). *Dampak listrik masuk desa dan perusahaan listrik non-PLN: Provinsi Sumatera Utara dan Sulawesi Selatan*. Jakarta: BPS.

Dakung, S., Sitanggang, H., Manan, F. N., Wahyuningsih, D., & Waluyo, H. (1990). *Dampak listrik masuk desa di Desa Cisadane, Keca- matan Cibadak Kabupaten Sukabumi*. Jakarta: Departemen Pendidikan dan Kebudayaan, Direktorat Jenderal Kebudayaan, Direktorat Sejarah dan Nilai Tradisional Proyek Inventarisasi dan Pembinaan Nilai-Nilai Budaya. Departemen Pendidikan dan Kebudayaan.

ESDM. (2020). *Statistik Ketenagalistrikan 2019*. Jakarta: Sekretariat Jenderal Ketenagalistrikan.

Gertler, P. J., Martinez, S., Premand, P., Rawlings, L. B., & Veermeersc, C. M. J. (2011). *Impact evaluation in practice*. The World Bank.

IEA. (2013). *World energy outlook 2013*. OECD/IEA.

Khandker, S. R., Barnes, D. F., & Samad, H. A. (2013). Welfare impacts of rural electrification: A panel data analysis from Vietnam. *Economic Development and Cultural Change, 61*(3), 659–692.

Laskari, F. (2012). Lisder Bukan Tanggung Jawab PLN Rayon Sungailiat. *Tribunnews.com*. Accessed on May 2013 from http://bangka.tribunnews.com/2012/07/30/lisdes-bukan-tanggung-jawab-pln-rayon-sungailiat

PLN. (2012). *Rencana usaha penyediaan tenaga listrik PT. PLN 2013–2022*. Jakarta; PT. PLN.

PLN. (2014). *Statistik PLN*. Jakarta: PT. PLN.

Republic of Indonesia. (2007). Energi. *Undang-undang Republik Indonesia No. 30 tentang Energi*.

Sambodo, M. T., Fuady, A. H., Negara, S. N., Handoyo, F., & Mychelisda, E. (2014). *Laporan penelitian model dan strategi pengembangan sektor ketenagalistrikan di daerah dalam upaya pengentasan kemiskinan*. Jakarta: Pusat Penelitian Kependudukan.

United Nations (UN). (2002). *Report of the world summit on sustainable development*. Johannesburg, South Africa, 26 August–4 September. Accessed on October 8, 2014 from www.unmillenniumproject.org/documents/131302_wssd_report_reissued.pdf

UNDP. (2005). *Energizing the millennium development goals: A guide to energy's role in reducing poverty*. UNDP.

Widodo, J., & Kalla, Y. (2014). Jalan perubahan untuk Indonesia yang berdaulat, mandiri dan berkepribadian: visi misi dan program aksi. Accessed on October 8, 2014 from http://kpu.go.id/koleksigambar/VISI_MISI_Jokowi-JK.pdf

Chapter 5
Epilogue: Agenda for Development of Rural Electrification

5.1 Introduction

Energy poverty, especially electricity, is a development challenge that requires an immediate solution. In 2019, more than 810,148 households in Indonesia, did not have access on electricity and 2,283,581 households having electricity, but not from PLN (ESDM, 2020a). Generally speaking, those households live in remote areas or they had access off grid supply with limited supply of power and high risk of interference. Although at the global level, Indonesia has contributes significantly to reduce energy poverty, electricity consumption in Indonesia is still relatively lower than other countries with similar income level or even lower than Indonesia.

Further, Indonesia is considered fruitful in improving the household access to modern cooking energy services. The energy conversion program from kerosene to Liquid Petroleum Gas (LPG) has successfully scaled up the penetration of LPG. In 2009, the share of LPG consumption to the total energy consumption at household level was about 28.7%, while kerosene was about 30% (ESDM, 2020b). However, in 2019, the share of LPG and Kerosene was about 48.87% and 2.21% respectively (ESDM, 2020b). This has been a very sharp rise of LPG consumption over the past two decades.

Unlike the great success in the kerosene to LPG conversion program, Indonesia has a different story in providing electricity access to the communities. Although the share of electricity consumption to the total energy consumption at household level had increase from 41.67% to about 48.61% between 2009 and 2019, the speed was lower than LPG (ESDM, 2020b). The government's efforts to improve access to electricity for the communities have not made a substantial progress. The accelerated power development program through the fast track phase I and phase II, which has been launched since 2006, was delayed due to various factors. As a result, many newly built power plants failed to reach their optimal operations. As seen from previous chapter, in 2010, 2012, and 2013, PLN could not serve optimally new connection request.

M. T. Sambodo et al., *Electricity Access and Community Welfare in Indonesia*, SpringerBriefs in Energy, https://doi.org/10.1007/978-981-16-3611-0_5

In the document 'PT PLN (Persero) Electricity Supply Business Plan, 2019–2028, it is stated that in 2020, all households in Indonesia will have access to electricity. If this target can be achieved, then Indonesia will be 10 years faster in achieving the SDGs. However, in the context of comparison with other countries, Indonesia's position is relatively late in achieving 100% of the electrification ratio (RE), for example Vietnam was able to achieve 100% RE in 2017 and China in 2010.

Although 100% RE can be achieved this year, it does not mean the problem is over. There are at least three things that need attention from stakeholders. First, as seen from previous chapter, households get access to electricity from various sources or are in the category of PLN customers and not PLN customers. In 2019, it was recorded that around 2.28 million households had access to electricity not from PLN or around 3% of the total number of households. Of that date, about 80% of the houses are outside Java Island. This fact has implications for energy justice, especially, especially in terms of access to electricity subsidies, under the COVID-19 situation, as part of social protection program.

PLN electricity customers, especially those who are included in the target group of subsidy recipients, namely 450 VA customers and poor customers at 900 VA power. In fact, due to COVID-19, the 450 VA customer group got free electricity and a 50% discount for 900 VA customers. Non-PLN electricity customers are certainly not as fortunate as PLN customers because they will not get electricity subsidies, even though from a household economic condition they are entitled to electricity subsidies.

Second, access to electricity does not reflect sufficient electricity conditions. Low level of electricity consumption can bring more challenge for more productive used of electricity.[1] Indonesia's per capita electricity consumption, it has only reached around 812 kWh. This value is far below Malaysia and Vietnam, which in 2014 had respectively reached 4652 kWh and 1424 kWh. In fact, in the standard figures provided by the World Energy Agency (International Energy Agency), the electricity consumption requirement per household is around 1250 kWh. The relatively low number of electricity consumption both per capita and also at the household level indicates that the benefits of the presence of electricity have not been fully optimized to support a better life at the household level. This can occur due to limited access to electricity and purchasing power.

Third, in several provinces, it was found that RE figures were quite high, such as Papua around 94.3%and East Nusa Tenggara 85.8%. However, if you examine the amount of RE not from PLN for the two provinces, it is about 45% and 30% respectively. This indicates the relatively high role of non-PLN actors.

Indonesia should make measurable and effective breakthrough efforts; otherwise, it will have difficulty in improving access to electricity for the people at all levels, especially those in remote rural areas. The demand for electricity will be even higher as the population and people's income are rising. Therefore, it is imperative for the government to speed up its efforts to increase the electrification ratio so that the target of 100% electrification ratio can be fulfilled by 2020 as planned.

---

[1] https://data.worldbank.org/indicator/EG.USE.ELEC.KH.PC (Access 6 January 2021).

This chapter is an attempt to establish policy recommendations for the advancement of rural electrification ratio based on the research findings and lesson learned during the 2013–2014 period. This chapter is divided into three parts. The first is to review several programs of electricity access improvement for the societies, especially in rural areas. The second is to analyze achievements and challenges of the programs that had or have been running. The last is to recommend steps that the government can take to increase access to electricity, especially in rural areas.

5.2 Evaluating the Electrification Programs

The Law No. 30/2009 on Electricity stipulates that electricity should be controlled by the state and provided by the central and regional governments under the principle of regional autonomy. This rule implies that the provision of electricity supply in practice should be first put in the hands of the State-Owned Enterprises (BUMN) and the Regional-Owned Enterprises (BUMD). Nevertheless, the RUKN 2013–2031 stated that private-owned companies, cooperatives and non-governmental organizations are allowed to participate in electricity supply businesses. The RUKN has an essence to safeguard the very significant roles of the central and regional governments in providing electricity, especially for the poor people living in the underdeveloped or in the border and remote areas.

According to the Ministry of Energy and Mineral Resources, in 2019 the ratio of electrified villages has reached around 99.48% of the 83,436 villages. In another word, there are still around 0.52% of Indonesian villages without access to electricity. The ratio of electrified villages is defined as the percentage of households that have access to electricity in a given village. That is, the electrification ratio in one village is considered 100% if there is one electrified household regardless of the fact that many others are still lacking access to electricity. The electrification ratio should better show the percentage of households with access to electricity compared to the all-existing households. Meanwhile, the government has aimed to attain an electrified village ratio at 99% by 2020.

As shown in Fig. 5.1, the central and regional governments and PLN play a key role in expanding access to electricity for the rural communities. The government is obliged to manage development in electricity using the State Budget (APBN) for the improvement of people's welfare as mandated by the Law No. 30/2007 on Energy and Law No. 30/2009 on Electricity. In performing the implementation of electricity development, the government has assigned PT PLN as the dominant operator to build and manage the power plants and their networks in a transparent manner. Currently, the government and the PLN have been exercising a number of cheap and economical electricity programs belonging to the central and regional governments; they have been also working together on PLN's on-grid electricity programs. The off-grid interaction patterns between the government and PLN have been taking place through the implementation of solar power plant (PLTS) programs, including decentralized solar power plants or the so-called solar home system (SHS), centralized solar power

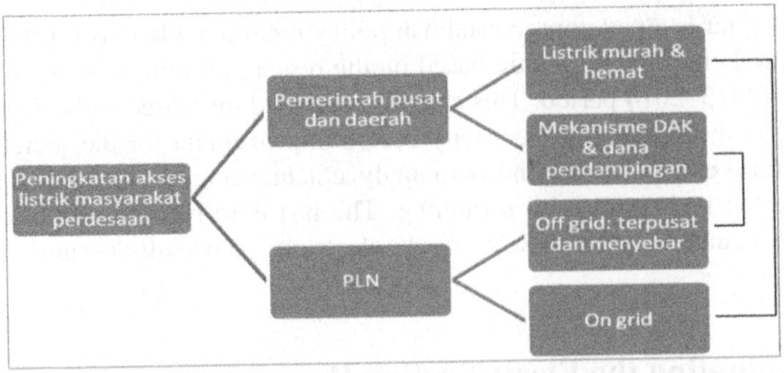

Fig. 5.1 General Scheme of Rural Electrification Program. *Note* Peningkatan akses listrik masyarakat pedesaan = Improvement of electricity access for rural communities; Pemerintah Pusat dan Daerah = The Central and Regional Governments; Listrik Murah & hemat = Cheap and Economical Electricity; Mekanisme DAK & Dana Pendampingan = Mechanism of DAK and assistance funds; Off-grid: terpusat dan menyebar = Off-grid: centralized and decentralized. *Source* Sambodo et al. (2014)

plants funded by the Special Allocation Fund (DAK), and decentralized solar power plants funded by the PLN (SEHEN). The following section will examine the extent to which the interaction patterns have been running so far.

The central government has been heavily reliant on the DAK mechanism to increase the rural electrification ratio. The Special Allocation Fund (DAK) is attained by compelling the district/city governments to provide assistance funds amounting to 10% of their allocation funds. The funds are used for physical developments based on the proposed activities submitted to the Ministry of Energy and Mineral Resources (ESDM) whose role is then to establish technical criteria and guidelines for the implementation of DAK-funded rural electricity programs (Ministry of Energy and Mineral Resources, 2014a). Meanwhile, the Mining and Energy Service at the district/city levels are responsible to execute the programs starting from planning and auction to monitoring and supervision.

The mechanism of Special Allocation Fund (DAK) taking place since 2011 has a difference from the previous one. Before 2011, all the processes of planning, auction, monitoring, and supervision were centralized in the Ministry of Energy and Mineral Resources (ESDM). The Ministry of ESDM has been running the programs of electricity access improvement for the communities through the Solar Power Plant (PLTS) program or the commonly-called Solar Home System (SHS), either decentralized (i.e. owned by each household) or centralized (or managed by the local community organizations). The SHS program has received special attention from the Ministry of Energy and Mineral Resources because it was reasonably fast and easy to build. The only challenge with this program was that most of its components must be imported and that the maintenance and repair facilities are still insufficient. Moreover, this centralized system of planning, auction, and monitoring is prone to fraud. The Corruption Eradication Commission (KPK) found that there was abuse of power at that time by the Directorate General of Electricity and Energy Utilization of the Ministry of Energy and Mineral Resources (Ministry of Energy and Mineral Resources, 2014b).

The absence of a centralized system in the current mechanism of Special Allocation Funds (DAK) is expected to bring more transparency into the program. No cases of corruption have been found so far related to the use of Special Allocation Fund (DAK) for rural electricity (later called rural energy). This is evidence that the Ministry of Energy and Mineral Resources has learned a lesson from their previous mistakes. Since 2011, the Ministry of Energy and Mineral Resources has issued technical guidance regarding the use of Special Allocation Fund for rural electrification. For example, the Ministry launched the Minister Regulation No. 3/2014 on Technical Guidance on the Use of Special Allocation Fund for Rural Energy, Budget Year 2014. According to the rule, the Special Allocation Fund for rural energy can be used for (i) micro-hydro development; (ii) rehabilitation of centralized micro-hydro and solar power plants; (iii) service improvement in centralized micro-hydro and solar power plants; (iv) establishment of centralized and decentralized solar power plants; and (v) construction of household-scale biogas installations (Ministry of Energy and Mineral Resources, 2014a). Nevertheless, the regulation does not mention the use of Special Allocation Fund for maintenance and reparation of the existing centralized solar power plants. Thus, the household users bear the responsibility to rehabilitate the decentralized solar power plants.

The Minister Regulation No. 3/2014 also asserts that the Special Allocation Fund for rural energy should be prioritized to villages unreachable by PLN electricity networks. To make sure that the use of Special Allocation Fund goes according to plan, the central government has allocated deconcentration funds for the provinces to monitor the recipients of Special Allocation Fund among the districts/cities. Monitoring and evaluation have been also strengthened by establishing a Regional Work Unit *(SKPD)* obliged to provide regular reports (quarterly and annually) to the Head of District *(Bupati)*. Henceforth, the Head of District will submit reports to the Minister of Energy and Mineral Resources, the Minister of Finance, the Minister of National Development Planning Agency, Minister of Home Affairs, with a copy to the Governor, the Secretary General of the Ministry of Energy and Mineral Resources, and the Director General of New, Renewable Energy, and Energy Conservation (EBTKE).

In addition to the mechanism of Special Allocation Fund (DAK), the government has been also undertaking a cheap and economical electricity program since 2012. The program aimed to help the poor and disadvantaged people get access to electricity. Under this program, the Target Households (RTS) received a free electricity installation. The Target Households are determined by three main criteria, i.e. (i) they are categorized as poor households; (ii) the houses belong to their own; and (iii) The distance from the houses to the PLN distribution point is no more than 40 m (Ministry of Energy and Mineral Resources, 2014a). Furthermore, the PLN, together with the team at the village and sub-district levels, and the Regional Development Planning Agency *(Bappeda)*, will verify the Target Households (Ministry of Energy and Mineral Resources, 2014a). However, not all The Target Households can have access to the cheap and economical electricity program. For example, in Nunukan

Sub-district, Nunukan District, North Kalimantan Province, there were 2000 Target Households, but only 500 had been given services according to the sub-district's quota (Mujayatno, 2013). The inexpensive and efficient electricity program is designed to integrate with PLN's electrification program.

As mentioned in the previous chapter, PLN uses two models of electricity access improvement for the communities, i.e. on-grid and off-grid. As for the on-grid program plan, PLN explains its work plan on rural electricity programs in the 2013-2022 RUPTL. As seen in Table 5.1, PLN seeks to have more than 2.8 million new customers in rural areas over the next ten years. The list of investment plan has included the Medium Voltage Network (JTM) and Low Voltage Network (JTR). Besides, more than 900,000 of the Target Households (or around 31% of the total new customers of rural electrification) will enjoy the inexpensive and efficient electricity program.

In addition to the on-grid electricity program, PLN has been also carrying out an off-grid electricity program called the Super Savings Energy (SEHEN), which is indeed not part of the development program of electricity system capacity (PLN, 2013). The interviews with PLN revealed that the SEHEN program is actually temporary and only instigated in several provinces with low electrification ratios, such as NTB, NTT, and Papua. In this case, it seems that the SEHEN program aims solely to pursue the target of electrification ratio being set by the government. According to the informants, PLN will replace the SEHEN program with a centralized or communal solar power generation, which is of course intended to increase the power capacity and to improve PLN's services for the communities.

5.3 Analyzing the Success and Failure of Rural Electricity Programs

The presence of on-grid and off-grid electricity programs means a lot to some people who have been waiting for decades to have electricity access that brings both monetary and non-monetary benefits. The financial advantages that people have after gaining access to electricity include lower kerosene expenditure and longer working hours, especially at night. The money saved from the reduced kerosene spending is reallocated to other needs, normally to increase household expenditure. Access to electricity also helps people save money from their energy expenditures. In some remote rural areas, most people are still using kerosene-fueled lamps or torches for lighting. After having access to electricity, they can now reduce their spending on kerosene fuel. Meanwhile, the non-monetary benefits bring a sense of safety among people because lighting at night makes their villages brighter and livelier. The children can study longer at night. In addition, the air quality in the house is much better than when using kerosene as a lighting fuel.

For most people who have access to electricity through on-grid programs, the main challenge is that the costs of connections and installations are rather expensive

Table 5.1 Development plan of rural electrification program in Indonesia, 2013–2022

Tahun	Unit ukur	2013[a]	2014	2015	2016	2017	2018	2019	2020	2021	2022	Total
JTM	kms	6.345	4.409	4.731	5.092	5.505	5.951	6.413	6.908	475	8.066	6.087
JTR	kms	4.736	4.412	479	5.143	5.529	599	645	6.938	7.524	8.149	5.966
Trafo	MVA	218	198	218	235	253	273	294	317	343	369	2.717
	Unit	3.446	3.217	3.488	3.763	405	4.371	4.718	5.083	5.503	5.954	43.593
Jml Pelanggan	PLG	22.017	214.493	2.304	247.317	268.127	290.605	311.672	334.214	361.961	392.871	2871.263
Listrik Murah & Hemat	RTS	95.227	951	88.889	88.889	88.889	88.889	88.889	88.889	88.889	88.889	901.438

[a]DIPA

Note Tahun = Year; Trafo = Transformer; Jml Pelanggan = Total Customers; Listrik Murah dan Hemat = the Inexpensive and Efficient Electricity

Source PLN (2013)

according to their income standards. Under this condition, they usually prefer to pay in installments. They will not make an electricity connection when they cannot afford it. In that case, they will usually get an electricity connection from the nearest neighbors. The government needs to push the efforts to expand the inexpensive and efficient electricity program to the poor people.

The lack of electricity supply is another problem in the field. It is true that on-grid electricity has been connected, but PLN has to make a regular blackout more than once a day due to the limited power supply. This instable electricity supply has encouraged some people to keep buying kerosene to anticipate the blackout.

In regard with the off-grid program, the government must pay attention to the synergy between the construction of decentralized solar power plants (SHS) and SEHEN. While the former is funded through the scheme of Special Allocation Fund (DAK), the latter is through the PLN budget. It is quite often that the two programs have clashes with each other, particularly regarding their very different aspects of management. The SHS program from the ESDM Ministry is classified as a grant to the communities for which they are exempt from all the charges. On the contrary, the SEHEN program from the PLN requires the customers to pay a monthly fee. The difference in terms of charges has created a social problem. A jealousy grew among the SEHEN customers (who are charged for monthly fees) of the households who received a free solar power generation from the government. There has been a sense of injustice because the poor people are treated differently. Therefore, there is a need to give much thought on the aspects of justice for the sustainability of the rural electricity program in the future.

Table 5.2 draws a comparison of differences between the solar power plant under the scheme of the Special Allocation Fund (DAK) and that of the PLN fund. Although both programs aim to help people get lighting, they have different managerial features leading to dissimilar costs and benefits for the communities. From a community perspective, the SHS program is better than the SEHEN because it offers social assistance. However, in term of program sustainability, the SEHEN program is more promising because it is managed by the PLN, which has more competence and capacity in the installation and maintenance of electricity system. The SHS program, on the other hand, is lack of maintenance and financial assistance for rehabilitation. If priority is given to the sustainability of the program, the SHS program needs to be evaluated by imposing all the maintenance costs upon the communities.

In addition to the major problem of sustainability, the involvement of other ministries in undertaking the SHS-based programs tends to create an overlap. In several places, the research team found some cases wherein people received the same SHS program assistance from different ministries. The lack of coordination between the ministries resulted in this overlap. It is common that people would request a program from other ministries after being denied by one ministry due to technical or non-technical reasons. This condition indicates that there is inefficient management in budgeting for rural electrification. In this case, the suppliers of SHS electrical equipment and the parties who see it as a mere commercial project become the most beneficiaries of the various SHS programs. Thus, the government should

Table 5.2 Major differences between the solar power plant (PLTS) Programs by the Ministry of ESDM and by PLN

Indication	Ministry of ESDM	PLN
Usage fee	Free of charge for the SHS; and applicable for the centralized PLTS, the amount of which depends on the agreement the users	A monthly usage fee is applicable for the SEHEN, amounting to Rp. 35,000, and for The centralized PLTS according to the social classification tariff, which is Rp. 18,400 per month
Parties in charge	The corresponding Department and Contractor during the guarantee period; and after the operation managed by the local community organization for the centralized PLTS and owned by each household for the SHS	PLN Sub-Branch
Technical capability	Limited	Major competence
Access to spare parts	People themselves should search for	PLN's responsibility
Payment method	Managed by the society (easy)	Requires transportation costs to get to the nearest bank or PLN office
Electricity power (per household)	50–100 WP	12 W (SEHEN) and varied up to 450 WP (PLTS)
Maintenance cost	Paid by the User	Paid by the PLN

Source Summarized by the author from different sources

be more serious in evaluating the quality, not only the quantity, of the sustainability of SHS programs.

In general, the current development of electricity access is taking an unsustainable pattern. There should be a serious effort to prepare well-trained human resources in order to address the sustainability issues. At present, the regional governments bear the responsibility to monitor and implement the program. They are lacking for competent human resources to handle technical and non-technical limitations. This shortage is particularly evident when the program is encountered with technical problems. As a result, the program is not going according to plan and its quality is gradually declining.

One of the field findings showed that the local community organizations were having difficulties in managing the centralized solar power plants from the Ministry of Energy and Mineral Resources. They complained for the lack of knowledge to deal with technical problems and to find necessary spare parts when the devices are broken. Although they have been given training to overcome technical problems—because their educational backgrounds and expertise are not in electricity—they are still having difficulties. Therefore, the quality of the program was low and the level of people's satisfaction with the program services was weakening.

In addressing this problem, the regional governments tried to allocate maintenance funds for the centralized solar power plants. However, because they are short of experts in this field, the level of reliance on suppliers was still high. Furthermore, the suppliers in general were not producers, rather retailers who sell the products so that it was hard for them to find the spare parts. This problem is inseparable from the fact that the local industries related to the components and spare parts of the solar power plants have been underdeveloped. Therefore, there is no wonder that the program of solar power plants is of high dependence on the imported components.

Based on the solar-based electricity programs undertaken so far in rural areas, there are at least four main factors contributing to the failure of the programs. The first is lack of synergy between the programs run by the ministries and the PLN. Each program has its different objectives, standards and procurement criteria. The second is reliance on imported equipment or spare parts due to the absence of supports from the local industries. The solar power plant-based programs require a large number of imported components for which affordable and high-quality spare parts are not easy to find. In addition, different technical standards and specifications imposed by each supplier make it more difficult for the people to get proper spare parts. The third is absence of competent human resources to overcome technical problems. The construction and maintenance of solar power plants need for special expertise because of many interrelated electronic devices. The rural communities are not skillful enough to deal with technical installation and maintenance. They still need for special training and guidance to properly operate and maintain the equipment of solar power plants.

As a result, the rural communities heavily depend on the program manager for the operationalization of the solar power plants. The fourth is unpreparedness of the supporting institutions to operate and finance the rural electrification for the poor people. PLN has stopped temporarily the SEHEN program due to high public arrears. The delayed payments happened because of various reasons, one of which is the relatively expensive cost of electricity subscription according to their income standards. The temporary suspension of the SEHEN program was mainly because PLN and its supporting organizations were unready to run this program sustainably. PLN argued that the SEHEN program has incurred financial loss due to high public arrears. This argument seems to be ironic because, as a State-Owned Enterprise (BUMN), PLN needs to take into account the best interests of the communities, especially the poor ones.

Based on the evaluation of various solar power plant programs by several ministries and PLN, it can be inferred that planning and budgeting systems in Indonesia are poorly linked. There are several ministries and institutions with budgets to run rural electrification programs but they, unfortunately, do not have sufficient competence in the electricity sector. PLN has indeed the competence but the demand more on profits than public interests has pushed it to remove rural electrification programs from its top priority. The temporary shutdown of the SEHEN program provides a good example of the weak institutional and organizational planning within the PLN. Thus, it is very important for the government and other stakeholders to work together in finding the best solutions to ensure the sustainability of rural electricity programs.

5.4 Institutional Reforms

Law No. 30/2007 on Energy and Law No. 30/2009 on Electricity underline the responsibility of the central and regional governments in securing access to energy, especially electricity, for the communities. The people's need for electricity access can be answered through various ways, such as on-grid and off-grid programs. The Laws also allow the involvement of business actors other than the State-Owned Enterprise (BUMN) and the Region-Owned Enterprise (BUMD). The private sectors, cooperatives, and non-governmental organizations can take part in the electricity development. The common challenge in rural electrification programs is how to expand electricity access to all socially and economically disadvantaged groups.

The PLN's on-grid programs have been limited to the community groups who can afford the installation and subscription fees. At the same time, some disadvantaged groups, who generally live in remote rural areas, have benefited only from off-grid electricity programs under the mechanism of Special Allocation Fund (DAK) and PLN budget, or under the participation of the private sectors. Nowadays, the inexpensive and efficient electricity program carried out by the government has not been able to serve all the disadvantaged groups. One of the reasons was that the high costs of electricity connection and installation has posed financial difficulties to some community groups in getting electricity connection. To overcome this problem, the regional governments need to set up a different standard of connection fees for the disadvantaged community groups. PLN, as the operator, can provide cross-subsidies to them. In addition, the local governments need to strengthen the role of savings and loan cooperatives or other microfinance institutions. The optimal operation of these institutions would help to reduce the connection costs for the rural communities.

As for the off-grid electricity programs, the government needs to have a clear mechanism in order to reach out the target groups. The government, PLN and other stakeholders should have a good communication with each other in order to create synergy between the rural electricity programs. At present, there are many overlaps between the programs of the ministries and those of the PLN. The government should eliminate sectorial arrogance that will endanger the sustainability of the programs and will cause a budget inefficiency. The most urgent step now is to ensure that the off-grid electricity programs are running through one-door coordination by, for instance, the Ministry of Energy and Mineral Resources. This one-door policy is important to minimize overlapping programs and to make sure that those programs adopt the same standards and evaluation criteria. Before running an off-grid electricity program, the first and foremost criterion is the sustainability of the program in all aspects, from planning, operation, maintenance, to rehabilitation. It is of importance to consider the issuance of government regulations that will provide a legal umbrella for the planning and implementation of off-grid electricity programs.

With reference to the decentralized SHS program, assets that have been built by the central government are, in principle, granted to the ownership of the regional governments. Thus, the regional governments are responsible for maintaining and maintaining the granted assets. However, for a sustainable management of the SHS, the district/city government needs assistance from the central government or PLN that has the necessary technical competence in the field. One consideration is to delegate the SHS assets to the village governments. The regional governments (*Pemda*) can encourage the establishment of Village-Owned Enterprises (BUMDes) to take care of the rural electricity business. The central and regional governments can invite experts from universities or research and development institutions to help the BUMDes deal with the technicalities of operation and maintenance. This mechanism will promise with certainty the sustainability of the SHS program in the future. In principle, sustainable rural electrification programs would require the strengthening of institution, human resources, and funding. The Ministry of Village, Development of Underdeveloped Areas, and Transmigration has the responsibility to improve village institutions and their human resources, and to provide adequate funding for the rural areas. Together with the Ministry of Energy and Mineral Resources, this ministry should ensure the development of basic infrastructures, especially electricity, in all the rural areas in Indonesia.

To boost the active participation of BUMN, BUMD, the private sectors, cooperatives, and non-governmental organizations in developing the electricity sector in rural areas, there is a need to design the electricity programs for long-term benefits. For this reason, it is necessary to think of the synergy between the development of basic infrastructures, including electricity, and that of industrial and tourism sectors in rural areas. Some disadvantaged community groups need to be given financial assistance enabling them to utilize electricity for lighting. The SEHEN program run by PLN has been successful in giving temporary access to lighting for the communities in areas beyond the outreach of on-grid electricity programs. In the future, the government, together with PLN, must be able to ensure the sustainability of the program by releasing the costumers from a burdensome monthly fee and a complicated payment procedure.

Looking at the already existing institutional and organizational scheme, it is in fact quite promising to develop the SEHEN program. The SEHEN customers already have bank accounts by which it would be easier for the government to supervise and distribute assistance programs for the targeted poor households. The scheme that has been developed in the SEHEN program can also facilitate the government to synergize different rural development programs. It is important to keep in mind that the subsidies should be properly given to the targeted poor households and should be always monitored to avoid misdirected distribution. People with financial capabilities are obliged to pay full monthly fees, in accordance with the principle of cross-subsidies applied by the PLN for small customers.

In the future, the management of the SEHEN program should begin to be decentralized to the regional governments. Since it has been managed by the central government through the PLN, the spirit of decentralization as mandated by the Electricity Law has not yet been fulfilled. The principle of decentralization requires the active participation of regional governments in operating the electricity energy. The only problem is that the regional governments have a varied degree of capacity, which, by implication, necessitates a more effective work format for the operation of electricity.

The final objective is not only to increase the electrification ratio, but also to ensure the sustainability of production, distribution and rehabilitation of the electricity energy.

The government has to adopt the principles of efficient and effective management of rural electrification. In this case, the stakeholders should give priority to off-grid electricity development. The SHS and SEHEN programs can be revitalized and improved in terms of their efficiency and effectiveness. In general, the two programs have been clearly focused on the community groups living in remote areas far from on-grid electricity distribution networks. Therefore, the government and the stakeholders need to formulate programs and integrate them into the smallest social institutions, i.e. villages.

From the aspect of power plant development, the experience of Manggarai District, NTT Province, showed that the central government, through the PLN, remains the main player in providing electricity to the remote areas, while the Regional Government (*Pemda*) takes a very minimum role. Therefore, there is a need in the future to have the regional government more actively engaged in the management of PLTP Ulumbu. The Regional Government of NTT and PLN have not agreed yet on share of ownership regarding PLTP Ulumbu. Similarly, the case of Kamojang geothermal power plant proved that it is uneasy and takes a longer time for the regional government to attain revenue sharing in geothermal utilization. Pertamina, as the operator, must reach a Break Event Point (BEP) before being able to deliver a share in the profits to the Regional Government of Garut.

The regional governments tend to become passive due to their limited experience and their lacked technical and operational skills in creating a link between the electricity sector and the economic development in their regions. They can actually play a more active role in inviting investors to their regions. This will bring a positive impact on their regional electricity demand. Furthermore, a greater demand for electricity will result in a greater electricity production, which it means a higher income for the operator and a larger shared revenue for the regional governments. For this, the regional governments need to be more proactive and adaptive to a long-term development vision.

The government must also consider how to reduce the dominant role of the central government, especially PLN, in the electricity business. To encourage a greater participation of the regional governments, the central government and all the related parties must find a fair and balanced format of cooperation. The format should necessarily sequence the optimization of the role of the local governments in developing the electricity sector. For example, the phasing can begin with the simplest function, i.e. the planning of regional electricity sector. This is important to build a coherent and integrated flow of electricity planning between the central and the regional governments. In order to properly attain the goal, it is essential to improve quantitatively and qualitatively the human resources at the regional government level.

In the subsequent stage, the regional governments can establish a Region-Owned Enterprise (BUMD) assigned to focus on the electricity sector, from supplying and generating to transmitting and distributing the primary energy. To perform this process, it is important that the central and regional governments have a realistic view of their capabilities. A joint venture model of cooperation with BUMN, BUMD, or the private sectors becomes an option to improve the technical and financing capacities.

5.5 Model Scenarios of Off-Grid Electricity

Figure 5.2 exhibits the currently ongoing off-grid electricity models of Solar Power Plant (PLTS). The experience in the last five years showed that the SEHEN and SHS programs have been quite successful in improving access to electricity, especially for the people in rural areas. Both of the programs directly addressed the target households living in the areas unreachable by the PLN's on-grid electricity. For PLN, the SEHEN is a transition program before people have access to electricity through the PLN's on-grid networks. Meanwhile, the SHS program of the Ministry of Energy and Mineral Resources (ESDM) aims at serving people who have not gained on-grid or off-grid electricity from the PLN. However, there is a collision between the two programs in practice due to lack of coordination. As a result, some groups of people

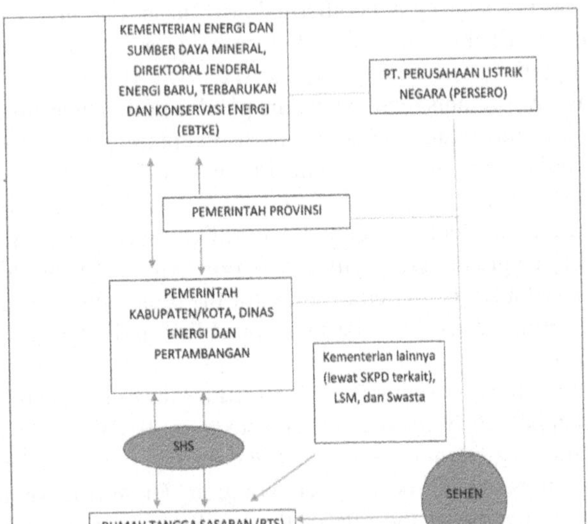

Fig. 5.2 Off-Grid model of solar power plant (PLTS) (A Current Scheme). *Note* Kementerian energy dan sumber daya mineral, Direktorat Jenderal Energi Baru, Terbarukan, dan Konservasi Energi (EBTKE) = Ministry of Energy and Mineral Resources, Directorate General of New, Renewable Energy, and Energy Conservation (EBTKE); Pemerintah Provinsi = The Provincial Government; Pemerintah Kabupaten/Kota, Dinas Energi dan Pertambangan = District/City Government, Dept. of Energy and Mining; Kementerian lainnya (lewat SKPD terkait), LSM, and Swasta = Other ministries (via the corresponding SKPD), NGOs, and Private Parties; PT. Perusahaan Listrik Negara (Persero) = State Electricity Enterprise (Persero); Rumah Tangga Sasaran (RTS) = The Target Households (RTS). *Source* Sambodo et al. (2014), modified

received both programs (SEHEN and SHS programs) in the same time, while some others got none. Meanwhile, rural electrification has been also carried out by other ministries through the Regional Work Units (SKPD) under the control of the related ministries. Non-governmental organizations (NGOs) and private sectors also help the communities improve access to electricity. The programs run by other than the government and PLN have a relatively small number of customers. There is also poor coordination between the stakeholders, wherein each party tends to operate within its own plan. In addition, the different costs and benefits of the programs create confusion and jealousy between the communities.

When it comes to management and equipment maintenance, the current model of the SEHEN and SHS programs are not yet sustainable. As mandated by Law No. 6/2014 on Villages, the current model should have given authority to the village governments not to only deal with the regulation and management of government affairs, but also with the interests of local communities, including the improvement of electricity access in rural areas. The availability of the law as a legal umbrella should be combined with operational technical capabilities for the optimum role of the village governments.

The current model of rural electrification development should adapt to the mandate of the new Village Law, which strives to make the village government as an entry point for all rural development activities, including the electricity sector. The active role of the village governments is expected to facilitate the implementation, monitoring and evaluation of the programs. The Village-Owned Enterprises (BUMDes) would be a key business entity in support of the implementation of rural electrification programs by the village governments.

We learn from the experience that the partnership between the Village Unit Cooperative (KUD) and the PLN had been quite well in the late 1980s. Table 5.3 shows the pattern of partnership that had run until the mid-1990s. However, nothing was heard about the continuity of the partnership after the reform era. The partnership pattern is potential to be continued because it had been successful enough in building the KUD's independence. It also had transformed the KUD from carrying out the very simple functions to becoming a self-sufficient business unit in the electricity sector.

Table 5.3 Patten of partnership between PLN and KUD

Pattern	Description
I	KUD reads the PLN meters, asks for the contribution fees, maintains the electricity networks, and deals with minor obstructions
II	In addition to taking the responsibilities as outlined in the pattern I, KUD can also carry out low voltage installations and maintain their networks as well
III	KUD buys bulk electricity from PLN or a private sector of the intermediate networks and then distributes it to the KUD members
IV	KUD runs commercial businesses in the electricity sector, ranging from generation, transmission to distribution, including installation to the KUD members

Source DPE (2000)

To ensure the sustainability of electricity access development for the communities, the central and regional governments need to strengthen the role of the Village-Owned Enterprise (BUMDes) which was established to improve the welfare of rural communities. The village governments will play a greater role if their financial capacities are improved through an increase in rural development funds. At present, the village revenues have been coming from two main sources: the Village Allocation Fund (ADD) and the Village Fund (DD) (Sambodo, 2014). The government has intended to divert all village-based activity budgets into village funds. The Government Regulation No. 43/2014 on the implementation of the Village Fund mentioned that around 70% of village funds must be allocated for poverty alleviation, education and health, infrastructure, and other programs that support the agricultural sector. In the future, it is crucial to integrate all the various ministerial programs of rural electrification in order to enlarge the Special Allocation Fund (DAK) for Rural Energy or to strengthen the village funds.

Corresponding to the spirit of the Village Law and the principle of autonomy and sustainability, the pattern of off-grid electricity access development in rural areas would be better under the management and scope of village organizations. There are at least two reasons for this. First, the village governments have been so far supporting the ministries and PLN in managing the SHS and SEHEN programs. For example, the village governments helped to collect the SEHEN subscription fees and submit them to the PLN. All maintenance issues remained the PLN's responsibility. Such a situation can be formalized into a more comprehensive framework of cooperation between the PLN and the village governments. In order to become a PLN's strategic partner in carrying out the SEHEN programs, for instance, the BUMDes needs to improve the capacity of its officials. The question now is whether the PLN will keep the SEHEN asset ownership or grant it to the village governments. The government, through the Ministry of Finance, the Ministry of Energy and Mineral Resources, the Ministry of State-Owned Enterprises (BUMN), and the PLN should sit together to discuss and provide the legal umbrella. Second, the local community organizations have been playing a vital role in supporting the centralized SHS and PLTS programs. Nevertheless, limited human resources had hindered them to become independent and well developed in managing the available assets of the solar power plants. Meanwhile, the management of SHS program has been delivered to each head of family.

In fact, the village governments have organizational and institutional functions that can be developed in support of the centralized Solar Power Plant (PLTS) and the SHS program. Under the scheme of Special Allocation Fund (DAK), the central government, which is in this case the Ministry of Energy and Mineral Resources, may transfer the PLTS and SHS assets to the district governments. Under the new Village Law, it is possible to bestow these assets to the village governments in which case they would be made responsible for the maintenance and operational costs. With the large amount of funding being allocated to villages, finance would not be problem in this case. However, it would require the village governments to have the necessary capacity in technicalities and procurement of components or spare parts. PLN, NGOs, or private sectors working on electricity generation efforts are expected to provide training or transfer of knowledge to the village officials.

Because financial and technical resources are distributed throughout various parties, it is important for the central and regional governments to safeguard the ability of the BUMDes to become a strong and reliable institution in handling the rural electrification programs. Thus, the process of transition for the transfer of assets from the PLN and the district governments to the village governments must be carried out measurably and gradually. The failure of rural electrification would reflect not only the shortfall of the government and PLN in increasing the electrification ratio, but also that of the village governments in improving the people's welfare. Thus, the village governments are the determinant actors in ensuring a smooth process of transition.

In addition to the need for coordination, the current model of institution needs to be upgraded in order to improve the performance of the apparatus and organizations. At this time, the implementation of the energy programs in rural areas are carried out by the Regional Work Unit (SKPD) at the district level. The SKPD is directly responsible to the Head of District who must report to the central government, represented by the Minister of Energy and Mineral Resources, the Minister of Finance, the Minister of BAPPENAS, and the Minister of Home Affairs. The report should also be copied to the Governor, the Secretary General of the Ministry of Energy and Mineral Resources and the Director General of EBTKE. With the deconcentration fund, the SKPD at the province level can conduct socialization, monitoring and evaluation of the SKPD at the district level. However, it turned out that many SKPDs at the district level are incompetent to carry out these functions. Because of the complex coordination and the lack of competence, it is necessary to improve the current model of institution.

Figure 5.3 recommends a model of off-grid electricity network management in rural areas. This new model has a unique feature in terms of strengthening the institutions of BUMDes as the frontline in the efforts to improve electricity access for rural communities. Under this model, all the rural electricity programs belonging to several ministries and institutions should be undertaken through one-door coordination of the Ministry of Energy and Mineral Resources (ESDM) that is responsible for improving access to electricity. The increase in the infrastructure budget at the Ministry of Energy and Mineral Resources can sustain the rural electricity programs. The Ministry of ESDM is also in charge of running the inexpensive and efficient electricity program through on-grid lines. In the face of more responsibilities and challenges, the Ministry of ESDM has no choice but to strengthen its human resources and institutional efficacy.

The provincial government serves as a facilitator between the central and the district governments. If in the past the flow of the Special Allocation Fund (DAK) had to stop at the level of district government, the new Village Law has now allowed the district government to delegate this authority to the village government. Thus, the implementation of the programs is entirely the duty of the village government, which is, of course, responsible to the district government. When running the programs, the village government can establish a BUMDes. It is needed that the village-based PLN's programs are included in the management of the BUMDes in order that the rural electricity programs are fairly and evenly distributed. Similarly, the participation of the private sectors and NGOs should be put within the framework of institutionally

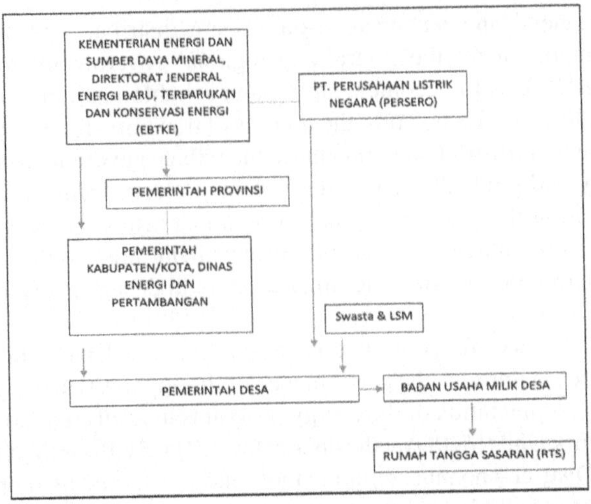

Fig. 5.3 Off-Grid Model with a More Active Role of the Village Government (LIPI's proposal). *Note* Kementerian energy dan sumber daya mineral, Direktorat Jenderal Energi Baru, Terbarukan, dan Konservasi Energi (EBTKE) = Ministry of Energy and Mineral Resources, Directorate General of New, Renewable Energy, and Energy Conservation (EBTKE); Pemerintah Provinsi = The Provincial Government; Pemerintah Kabupaten/Kota, Dinas Energi dan Pertambangan = District/City Government, Dept. of Energy and Mining; Swasta & LSM = the Private Parties and NGOs; Pemerintah Desa = The Village Government; Badan Usaha Milik Desa = The Village-Owned Enterprise; Rumah Tangga Sasaran (RTS) = The Target Households (RTS). *Source* Sambodo et al. (2014), modified

strengthening the BUMDes, rather than competing with it. Cooperation for mutual benefits should be developed among all the stakeholders. Furthermore, BUMDes is assigned to determine the allocation of the target households. The PLN, together with the regional governments at the district and village levels, plays an important role in improving the capacity of the BUMDes in rural electricity management. The pattern of cooperation between PLN and BUMDes can go through the stages as shown in Table 5.3. Referring to the model as recommended by the research team, it is necessary to transform the programs that have been currently undertaken. In the proposed model, the terms SEHEN and SHN are eliminated. Under one-door management, there will be uniform standards, criteria, and quality of programs.

5.6 Electricity Development and Poverty Reduction

The development of rural electrification should not stop at the lighting phase. It is of importance to think about how electricity development could affect the productive sectors to improve the people's income and quality of life. The electricity development in the District of Manggarai seemed to be solely oriented towards increasing electrification rather than developing the productive sectors. The existing electricity programs have benefited the communities in terms of lighting only.

Our observations found that the region contains many economic potentials that can be developed by the utilization of electricity energy. On a small scale, some households are already manufacturing products made from wood, such as cabinets and windows. Access to electricity will help them increase productivity and quality of work by using cutting or shaving machines. The research team also found that the coffee in the area can be further processed into featured products of the District of Satarmese. Thanks to the electricity access, electrical coffee dryers may be built for the coffee farmers. Furthermore, the electricity energy can also give additional value to the agricultural products, such as chocolate, palm sugar, and others.

The electricity sector development with a focus on empowering the society's productive sectors should consider the potential of available resources and institutions. Particular attention needs to be paid to whether the electricity developments are located at the center of energy resources, such as Ulumbu, or in the areas simply for electricity access provision. For the former, there would be more options or potentials to develop electricity in favor of neighborhood empowerment. For example, the development of hot spring tourism around the PLTP Kamojang and Darajat, West Java, is an option worth considering. The PLTP can be also developed into a destination for educational tours for the communities. The development of hydropower plants in Wonogiri shows a similar pattern, where the Reservoir of Gajah Mungkur becomes one of the main tourist attractions in the District of Wonogiri, in addition to being a source of agricultural irrigation in Central Java. It is obvious that the power plant-based development of tourism should be well regulated in order to prevent conflicts with that of electricity generation.

The economic development of the neighborhood around the power plants can be undertaken using the Corporate Social Responsibility (CSR) funds from the operators, such as Chevron for the PLTP Darajat and Pertamina Geothermal for the PLTP Kamojang. These CSR funds can be utilized to improve the community welfare aligned to its potential, such as cattle fattening or mushroom cultivation.

The development of alternative energy in Pantai Baru, Pandansimo, District of Bantul, is another example of electricity development oriented towards local businesses empowerment. This project was carried out independently by LAPAN, UGM, Kemenristek, KKP, and the District Government of Bantul. The alternative energy from wind energy, centralized solar panels and biomass has been developed in the region. The electrical power generated from this alternative energy has been used for lighting food stalls on the beach. From a cost–benefit analysis, the electricity sales from the project may not bring economic profits. However, the project has been successfully tapping the rich tourism potential in the region. Most people who used to depend on sand-mining have now turned to the tourism service sector. The windmills that adorn Pantai Baru become a special magnetism for the tourists to visit the region. In addition, Pandansimo has now become one of the main references for the development of alternative energy.

The electricity development in the areas far from the primary energy resource for electricity generation would depend on how much access to electricity can be achieved and the potential can be developed. If the installed power produces a limited amount of power to each household, the reasonable choice is to develop home industries, such as cake baking or convection. On the contrary, if the access to electricity was large enough to drive industrial machineries, the industrial development would have much more options, depending on the potential available in the areas.

The District of Manggarai, which is home to the PLTP Ulumbu and the potential for hydropower development, has certainly a considerable choice. Like in East Java, The Ulumbu geothermal field has the direct utility for hot water tourism like the one in West Java. The unused geothermal at the PLTP can also help the farmers make *moke* (a traditional beverage of this region) or distill oils from the clove leaves that have been currently underused. The development of the economic potential around the area is still highly possible and more widely open with the availability of access to electricity.

References

Department of Mining and Energy (DPE). (2000). *55 years of mining and energy development*. Departemen Pertambangan dan Energi.
ESDM. (2020a). *Statistik Ketenagalistrikan 2019*. Direktorat Jenderal Ketenagalistrikan
ESDM. (2020b). *Handbook of energy & economic statistics of Indonesia 2019*. Kementerian Energi dan Sumber Daya Mineral.
International Energy Agency (IEA). (2014). *World energy outlook 2014*. IEA Publications.
International Energy Agency (IEA). (2019). *Energy security in ASEAN +6*. IEA Publications.
Ministry of ESDM. (2013). *Draft general plan of electricity (RUKN) 2012–2031*. Accessed on November 1, 2014 from www.energy-indonesia.com/02electrcitylaw/0130213RUKN.pdf
Ministry of ESDM. (2014a). The Minister Regulation No. 3/2014 on Technical Guidance on the Use of Special Allocation Fund for Rural Energy Budget Year 2014. Accessed on November 2, 2014 from http://prokum.esdm.go.id/permen/2014/Permen%20ESDM%2003%202014.pdf
Ministry of ESDM. (2014b). DAK untuk listrik perdesaan. Accessed on November 2, 2014 from http://esdm.go.id/news-archives/323-energi-baru-dan-terbarukan/5265-dak-untuk-listrik-perdesaan.html.
Mulyadi, A. (2011, Juli 15). KPK tahan pejabat ESDM. *Kompas.com*. Accessed on November 13, 2014 from http://nasional.kompas.com/read/2011/07/15/2015544/KPK.Tahan.Pejabat.ESDM
Minister of Energy and Mineral Resources of the Republic of Indonesia. (2014). The Minister Regulation of Energy and Mineral Resources No. 3/2014 on Technical Guidance on the Use of Special Allocation Fund for Rural Energy Budget Year 2014.
Mujayatno, A. (2013). Pemerintah gratiskan listrik 500 KK miskin Nunukan. Accessed on November 1, 2014 from http://www.kaltaraprov.go.id/news-view/465/pemerintah.gratiskan.listrik.500.kk.miskin.nunukan.html
PLN. (2013). *Rencana Usaha Penyediaan Tenaga Listrik PT PLN (Persero) 2013–2022*. PT. PLN.
PLN. (2014). *PLN statistik 2013*. PT. PLN.
Republic of Indonesia. (2007). Energi. *Undang-Undang Republik Indonesia No. 30 tentang Energi*.
Republic of Indonesia. (2009). Ketenagalistrikan. *Undang-Undang Republik Indonesia No. 30 tentang Ketenagalistrikan*.
Republic of Indonesia. (2014a). The Government Regulation No. 43/2014 on the implementation of the Village Fund.

Republic of Indonesia. (2014b). Desa. *Undang-Undang Republik Indonesia No. 6 tentang Desa.*
Sambodo, M. T., Fuady, A. H., Negara, S. N., Handoyo, F., & Mychelisda, E. (2014). *Laporan Penelitian Model dan Strategi Pengembangan Sektor Ke- tenagalistrikan di Daerah dalam Upaya Pengentasan Kemiskinan.* Jakarta: Pusat Penelitian Kependudukan.

Index

M. T. Sambodo et al., *Electricity Access and Community Welfare in Indonesia*, SpringerBriefs in Energy, https://doi.org/10.1007/978-981-16-3611-0

The manufacturer's authorised representative in the EU is Springer Nature Customer Service Centre GmbH, Europaplatz 3, 69115 Heidelberg, Germany. If you have any concerns regarding our products, please contact ProductSafety@springernature.com

Printed and bound by CPI Group (UK) Ltd, Croydon, CR0 4YY
02/10/2026
02226885-0003